Stellar Thoughts

星 思 维

第八届"星艺杯"设计大赛获奖作品集

星艺装饰文化传媒中心 编著

暨南大学出版社
JINAN UNIVERSITY PRESS

中国·广州

图书在版编目（CIP）数据

星　思维：第八届"星艺杯"设计大赛获奖作品集 / 星艺装饰文化传媒中心
编著. —广州：暨南大学出版社，2020.11
　ISBN 978 – 7 – 5668 – 2968 – 9

Ⅰ . ①星…　Ⅱ . ①星…　Ⅲ . ①建筑设计—作品集—中国—现代　Ⅳ . ①TU206

中国版本图书馆 CIP 数据核字（2020）第 178069 号

星　思维：第八届"星艺杯"设计大赛获奖作品集
XING SIWEI：DIBAJIE "XINGYIBEI" SHEJI DASAI HUOJIANG ZUOPINJI
编著者：星艺装饰文化传媒中心
………………………………………………………………………………………………

出　版　人：张晋升
策划编辑：杜小陆
责任编辑：黄志波
责任校对：刘舜怡　王燕丽　黄亦秋
责任印制：汤慧君　周一丹

出版发行：暨南大学出版社（510630）
电　　话：总编室（8620）85221601
　　　　　营销部（8620）85225284　85228291　85228292　85226712
传　　真：（8620）85221583（办公室）　85223774（营销部）
网　　址：http://www.jnupress.com
排　　版：广州良弓广告有限公司
印　　刷：深圳市新联美术印刷有限公司
开　　本：889mm×1194mm　1/12
印　　张：19.5
字　　数：250 千
版　　次：2020 年 11 月第 1 版
印　　次：2020 年 11 月第 1 次
定　　价：188.00 元

Create
Happiness
And
Deliver Joy

设计幸福　播种快乐

1 住宅·工程实景作品

Residence-Engineering Live-Scene Works

903 黄宅

903 Huang's Residence

让人感到轻松愉悦的家，是能够让居住者平衡独处时间和家庭时间的空间。

为了实现空间平衡，本案设计师为业主逐一调整了原有的空间布局，并规划了生活动线，顺应机能产生不同的空间向性。从入口处拉出直白明了的空间动线，将客餐区一分为二，以踏步台阶为界，保持空间的秩序感。

A home that makes people feel relaxed and joyful is a space that allows residents to balance their alone time with family time.

In order to achieve a spatial balance, the designer of this scenario adjusted the original spatial layout one by one for the owner and planned a new one according to the living habit to adapt to the function to produce different spatial orientations. A straightforward spatial movement is drawn from the entrance, and the dining area is divided into two parts, bounded by steps, to maintain a sense of order in the space.

项目名称：903 黄宅
项目设计：广东星艺装饰集团
项目地址：广东广州
设计师：吴家春

　　在设计中，让客厅腾出更大范围的利用空间，使格局得到最大限度的释放。回合式踏步形式设计表现出空间的"体块"感，使空间更显张力；介于公私领域间的多功能室，以推拉门的形式变化提供弹性开放或私密空间的选择。进入主卧，以咖啡色和白色为主色调的设计与整体空间相呼应，温厚又清爽。在床侧设计一处可席地而坐的休闲区间，凸显悠然的自在感。

In the design, living room vacates a wider space for use, so that the layout can be extended to the greatest extent. The design of the round stepping stairs expresses a sense of "bulk" in the space, which makes the space more tense; the multifunctional room between the public and private areas, employs the sliding doors, providing a flexible choice of open or private space. Entering the master bedroom, the design with brown and white as the main color echoes the overall space, warm and refreshing. A leisure area on the side of the bed where one can sit on, highlights the leisurely sense of freedom.

　　整个空间以不同的材质和色块堆叠，线性和量体配合，由公至私，以简驭繁，从暗至明之层次，承载着简练而富有层次的空间美感，贴合业主偏爱的现代风格抒发情调，以拒绝粗糙、追求高质感生活的态度精心打造每一处空间。

With different materials and color blocks stacked, the entire space is mixed with linear and volumetric segments, from public to private, using less to accomplish more, from dark to bright, carrying a concise and layered space beauty, which fits the modern style preferred by the owner to express the taste and carefully creates every space with the attitude of rejecting roughness and pursuing high-quality life.

山水庭苑

Lake Villa

在本案设计前，业主要求空间应呈现出利落、大气的感觉。基于业主是五口之家，家里有三个小孩，所以设计师在构建空间时，提出要把一家人的情感互动"调动"起来，交互设计，从而营造家庭的温馨感。以客餐厅为交互中心，有选择地以开放式为要义，设计出开放式的厨房、客厅、餐厅以及书房，充满张力的布局构建交流的界面，从而形成家庭起居不同深度的社交空间。

Before the design of this scenario, the owner requested that the space should present a concise and elegant feeling. Based on the fact that the owner is a family of five and there are three children in the family, when constructing the space, the designer proposed to "mobilize" the emotional interaction of the family members, using this kind of interaction design to create a sense of family warmth. Taking the guest dining room as the interaction center, and selectively taking openness as the main point, the designer aims to design an open kitchen as well as open living rooms, dining rooms and study rooms, therefore creating a communication interface full with tension and forming a social space with different depths of family living.

项目名称：山水庭苑
项目设计：广东星艺装饰集团
项目地址：广东广州
设计师：夏芳芳

弗朗明戈

Flamenco

本案采用金属色泽的线条把空间勾勒出多重层次感，搭配精致的水晶吊灯和描金花纹，显得美轮美奂。典雅厚重的沙发、金碧辉煌的桌椅与深色的柜子相得益彰，与精美的地毯一同衬托出优雅奢贵的气质，以繁见美，华丽内敛。

This scenario uses lines in metal color to outline multiple layers of the space, with exquisite crystal chandelier and gold pattern, looking splendid and magnificent. Elegant and thick sofas, gilded tables and chairs, and dark cabinets complement each other. Together with the exquisite carpets, they set off an elegant and luxurious temperament, which manifests a complex beauty, resplendent as well as low key.

项目名称：弗朗明戈
项目设计：广东星艺装饰集团
项目地址：广西南宁
设计师：黄林妮

800 年新基古村老宅

800-year-old Residence in Xinji Ancient Village

BEFORE

AFTER

本案是由南塘祖祠堂改建而成的，位于南塘祖艺术馆旁。

原建筑形体是一个老民居，似乎还能透过空间"看"到老人的生活印记。一楼一屋一小院，对于空间本体而言，缺乏些许娱乐性，缺少对年轻人的吸引力，亦缺失新住民所希望的体验感。

This scenario was reconstructed from Nantang Zu Ancestral Hall, located next to Nantang Zu Art Museum. The original building is an old residence, and it seems that the moments of the elderly can be "observed" through the space. One building, one room and one small courtyard, as far as the space itself is concerned, it lacks a little entertainment and the attraction for young people as well as lacks the sense of experience that new residents desired.

项目名称：800 年新基古村老宅
项目设计：广东星艺装饰集团
项目地址：广东广州
设计师：陈文辉

BEFORE

AFTER

BEFORE

AFTER

空间的合理性是首要调整方向。本案设计目标是 3 间住客套房，南面需要调整墙体位置以改变各空间的内部尺度。一楼原副体建筑希望改建成为一个客房，但 2 300 毫米的横向跨度完全无法达到希望的舒适尺度，圈梁的原始结构非常考验设计功力。设计师想到用工字钢重新做结构加固，把原有的楼板和圈梁用钢体支撑，然后拆除旧的承重红砖墙体，达到外扩的目的。

A proper layout is the primary concern. This scenario is oriented to 3 guest suites and the position of the wall body on the south side needs to be adjusted to change the internal scale of each space. The original auxiliary building on the first floor was proposed to be converted into a guest room, but the 2 300 mm horizontal span is completely unable to achieve the desired comfortable scale. The original structure of the ring beam is a challenge of design skills. The designer got the idea of rebuilding the structure with I-beams, supporting the original floor slab and ring beams with steel, and then removing the old load-bearing red brick wall to achieve the purpose of external expansion.

主体建筑的窗体非常小，设计师希望让每个空间都能看到院子里的杨桃树，让建筑、树和人融合在一起，所以用同样的设计手法对主体建筑靠院子的墙体进行支撑加固。为了增加景流量，在扩大的窗体中不考虑通风的窗体，而是在隔离200毫米处另做安排。

二楼的露天天台与小院在功能上是重复的，可以设计为一个公共的聚会场所。副体建筑外扩会对院子里的杨桃树造成破坏，为了减少对树体斜枝的损伤，外扩建筑形态依托钢架沿树体形态斜面而走，以树养形。

The windows of the main building are very small. The designer hopes that the carambola tree in the yard can be overlooked from every space, creating harmonious picture of buildings, trees and people. Therefore, the same design method is used to reinforce the wall structure of the main building near the yard. In order to increase the flow of scenery, ventilated windows are not considered in the enlarged windows, instead, they are arranged to another position which is 200 mm away.

The open-air roof and courtyard on the second floor are functionally repeated and can be designed as a public meeting place. The external expansion of the auxiliary building will cause damage to the carambola trees in the yard. In order to reduce the damage to the oblique branches of the tree, the external expansion of the building relies on the steel frame to coordinate the shape of the tree.

闫宅

Yan's Residence

客厅的落地窗承接阳光的洗礼，以"白灰木"的全屋搭色，让客厅显得更加宽敞明亮。采用无主灯设计，利用点状光源和线型灯带来照明。一方面，让光线分布更加均匀，避免了使用主灯带来的金碧辉煌的效果，营造出低调内敛的空间气质；另一方面，线型灯带独具线条美感和立体感，可以带来不一样的视觉感受。

The floor-to-ceiling windows of the living room accept the baptism of the sun, and the color of "white ash wood" is used to make the living room more spacious and bright. It adopts no key light design and uses point light sources and linear lights to produce light. On the one hand, it makes the light distribution more even, avoiding the magnificent effect brought by the use of the main lamp, and creating a low-key and restrained spatial temperament; on the other hand, the linear light strip has a unique line beauty and three-dimensional sense, which can bring about a different visual experience.

项目名称：闫宅
项目设计：广东星艺装饰集团
项目地址：广东广州
设计师：吴佳

  沙发墙面以灰蓝色墙漆为主，右侧柱子特意留白，在柱子上方开一个 2 厘米的小槽镶入黑色金属条，与入户玄关处的金属廊栅相呼应，为空间营造一种连贯性、延续性，同时增加现代感。

  餐厅延续了客厅的整体效果，同时又展现出独特的创意。餐桌由水吧与一整块胡桃木衔接而成，不仅可以就餐，还可以办公。餐桌岛台一体，黑白灰系与木饰贯穿整个空间。MELT 吊灯的灵感来自融化的玻璃，通过镜面抛光和真空金属化，塑造不规则的形态，如同融化的冰川内部、炽热的岩浆或者深空的图案。

The sofa wall is dominated by gray-blue wall paint. The column on the right is left blank. A small 2 cm slot is grooved on the top of the column for inserting a black metal strip, which echoes the metal railings of corridor at the entrance of the house, creating a sense of continuity in space as well as increasing the sense of modernity.

The dining room extends over the overall effect of the living room while showing unique creativity. The dining table is made up of a water bar and a whole piece of walnut wood, not only for dining, but also for office work. The dining table and island table are integrated, with black, white, gray and wooden decorations throughout the space. The MELT chandelier is inspired by molten glass. Through mirror finish and vacuum metallization, irregular forms are shaped like the interior of a melting glacier, hot magma or deep space patterns.

卧室以灰色调为主，让业主在此空间能放松神经，专注于此刻的宁静，与自己内心对话。深灰色软包背景墙使空间变得柔和，实木床头柜配上精致的金色小吊灯，简洁大气，单纯不烦琐，营造出宁静、优雅的睡眠空间。

The bedroom is dominated by gray tones, allowing the owner to relax and enjoy the peaceful moment, searching for the inner self. The dark gray soft-covered background wall softens the space, and the solid wood beside cabinet with exquisite golden chandelier is concise and atmospheric, simple and not cumbersome, creating a peaceful and elegant sleeping space.

晨光栖语

Morning Light Lyrics

玄关采用黑、白作为主色调，像是将黑暗的空间撬起一条缝，让光照进来，看似简单，却充分显示出它的作用。踏过玄关，完成短暂的归家仪式。

The entrance uses black and white as the dominant tone, like prying a gap in the dark space for a beam of light. It seems simple, but it fully shows its effect. On stepping through the entrance, a short homecoming ceremony is completed.

项目名称：晨光栖语
项目设计：广东星艺装饰集团
项目地址：广东广州
设计师：林婷婷

窗日

Sun out of Window

叹息西窗过隙驹，微阳初至日光舒。

在时间面前，没有颜色，没有形体，更没有装饰，只有西窗伸入的阳光触角，是真是纯，是永恒的絮语。

本案在格局上去除多余的墙体，让自然的光和风在室内自由流通。去掉多余的装饰和色彩，大面积的白色让空间成为纯净安谧的容器，给予居住者最大限度的心理包容。木，具有生发、条达的特性，是最具生命力的。木色给空间增添了一份温暖和生机。

Slightly sigh for the passing time by the west window, the sunshine from the rising sun feels so comfortable.

In front of the time, there is no color, no shape, and no decoration, only the tentacles of sunlight extending through the west window, presents the sense of pure and everlasting.

In this scenario, redundant walls are removed from the layout, allowing natural light and wind to circulate freely in the room. Excessive decoration and color are also removed, and the large white area makes the space pure and tranquil, giving the residents maximum psychological tolerance. Wood, with the characteristics of growth and streak, is the most vital. The wood color adds warmth and vitality to the space.

项目名称：窗日
项目设计：广东星艺装饰集团
项目地址：广东广州
设计师：彭文博

童行

Joyful Childhood

给孩子最好的礼物，莫过于一个美好的童年。与孩子"童"行，但愿我们能够像孩子一样，保持热忱的心去面对世界，去相信，去发现，去忘我。美好的生活在家体现为既能享受独处的悠闲，也能和最亲的人聚在一起，用心一步一步去丈量幸福。

The best gift for children is a wonderful childhood. Spending a joyful "childhood" with children, it is hoped that we can face the world with enthusiasm like children, to believe, to discover, and to forget ourselves. A wonderful life at home is embodied as being able to not only enjoy the leisure of being alone, but also be able to gather with the closest relatives and measure happiness by heart.

项目名称：童行
项目设计：广东星艺装饰集团
项目地址：广东广州
设计师：柳军友

与友"童"行，荣华富贵如云烟，再多的功名利禄都抵不过一个真正的知心朋友。约三五知己在家中品茶小酌，畅谈人生，乐享生活。中西厨的设定，让厨房不再作为单独的烹饪场所，而是情感交流的天地。用餐完毕，步出阳台，远眺江景，近赏盆栽，打造属于自己的都市绿洲，这样惬意的日子无论什么时候都值得期待。

Spending a joyful "childhood" with friends, wealth and rank are just like clouds. No matter how much fame and fortune we have, it is not worthy of a true close friend. Sit home with several friends over a nice cup of tea, talking about and enjoying life. The setting of Chinese and Western kitchens makes the kitchen no longer a separate cooking place, but a place for emotional interaction. After the meal, step out of the balcony, overlooking the river view, enjoying the potted plants like creating your own urban oasis. Such a pleasant feeling is worth expectation at anytime.

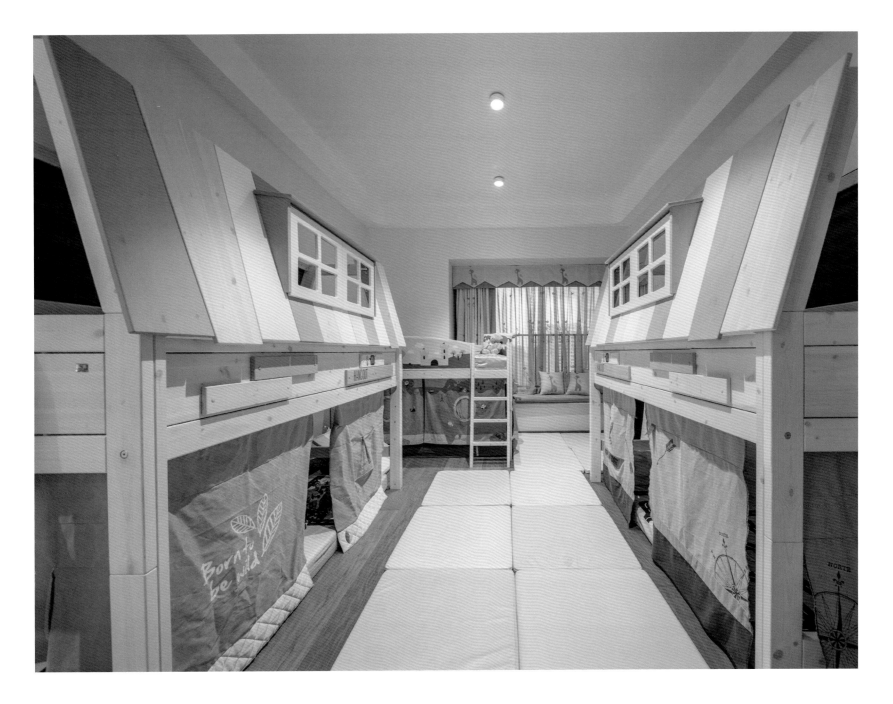

　　卧室是憩息的港湾。无须过多烦琐的装饰，嵌入式衣橱加上几盏调皮的灯，低饱和的色调，柔和的面料肌理，营造出静谧的生活氛围。长形洗手盆、独立卫生间、树屋房等专为小朋友们而设，让他们既能一同玩耍，也有独立的空间。

The bedroom is a haven for rest. Without too much cumbersome decoration, the embedded wardrobe plus a few nifty lamps, low-saturated tones, and soft fabric texture create a quiet living atmosphere. Long wash basins, independent toilets, tree house rooms, etc. are specially designed for children, allowing them to play together as well as owning an independent space.

归·心

Return to the Heart

项目名称：归·心
项目设计：广东星艺装饰集团
项目地址：广东广州
设计师：胡哲

大道至简，至繁终归于至简，至简助你理清初心，或者说回归本初，回归满足基本的功能。

Simplicity is the highest ideal and the complexity will eventually return to the simplicity. It will help you sort out your original mind, in other words, return to the beginning to satisfy basic requsets for functions.

　　本案设计注重空间上的功能优化，以满足使用者的功能需求。户型改造前存在采光较差、储藏空间不足等问题。为了增强空间的通透性，把阳台改为开放式的餐厨区，从而解决主卧小、卫生间出入不方便、储藏空间不足等问题。开放式厨房的设计增加了人与人之间的互动，它能够使人停留、互动交流，在这里可以喝茶、看书、聊天……厨房的地位上升为"心灵交流的屋"。

This scenario focuses on the optimization of functions in space to meet the functional needs of users. Before the renovation, there were problems such as poor lighting and insufficient storage space. In order to extend the space, the balcony was changed to an open kitchen, so as to solve the problems like small master bedroom, inconvenient access to the bathroom, and insufficient storage space. The design of the open kitchen increases the interaction between people. It creates a room for communication, where people can drink tea, read books, and chat... The status of the kitchen has risen to "a house of spiritual communication".

　　原卧室靠床位置的顶上有根凸出来的边梁，设计师在床头背景设计了一整排柜子，不仅让卧室的储物空间实现最大化，还完美解决了边梁凸出这个颜值硬伤。利用原来的卫生间区域腾出大部分空间给主卧，提升卧室的通风性；同时，把原来的小窗户改大，增加卧室的采光。墙角的树桩上放有一盆精致的植物，下午的阳光透过百叶窗照射进来，在墙上形成光影，让房间充满诗意与温馨。光是自然界最美的馈赠，让世界有了色彩，赋予空间以生命，让居住者获得使用空间的满足。

There was a protruding side beam above the bed in the original bedroom. The designer designs a row of cabinets on the bedside background, which not only maximizes the storage space of the bedroom, but also perfectly solves the ugliness resulting from the protruding side beam. The most part of the original bathroom area is changed to the master bedroom to enhance the ventilation of the bedroom; at the same time, the original small windows are enlarged to increase the daylighting of the bedroom. There is a pot of delicate plants on the tree stump in the corner. The afternoon sunshines comes through the blinds, forming light and shadow on the wall, making the room full of poetic harmony and warmth. Light is the most beautiful gift of nature, giving the world color, endowing the space with vitality, and satisfying the residents when they use the space.

生态城金府

Jin's Mansion in Eco-city

看似随意与轻松的设计，却能让人细细品味空间传递出来的文化内涵，感受它们为生活带来的温暖与感动。业主家里成员较多，三代同堂。结合别墅原结构以及业主的居住需求，别墅部分改建较多。改建后，一层为休闲区，二层为生活区，三层为私人区。客厅和餐厅的设计风格为欧式，在会客厅的设计中解决了采光问题，大气且奢华之感十足。

项目名称：生态城金府
项目设计：广东星艺装饰集团
项目地址：贵州贵阳
设计师：金星源

The seemingly easy and casual design allows people to savor the cultural connotations conveyed by the space and feel the warmth and touch they bring to life. There are many members in the owner's family consisting of three generations. In combination with the original structure of the villa and the residential needs of the owner, part of the villa was renovated. After renovation, the first floor becomes a leisure area, the second floor a living area, and the third floor a private area. The style of the living room and dining room is European style. The lighting problem is solved in the design of the reception room, which reveals a sense of grandeur and luxury.

随着生活品质的提升，单一的装修风格已经不能完全满足居住者的需求，为了突出个性和迎合居住者的品位，且考虑到不同年龄有不同的居住需求，设计师将不同风格中的优秀元素融合，打造出以舒适机能为导向的空间。设计师根据不同的需求，为每个居住者定制了专属于自己的私密空间。

With the improvement of the quality of life, the needs of the residents cannot be fully satisfied through a single decoration style. In order to highlight the individuality and cater to the taste of the residents as well as take into account of the different living needs of different ages, the designer manages to integrate the outstanding elements of different styles to create a function-oriented room.According to different needs, the designer has customized a private space catering to each resident.

主卧贴合自然气息，设计以安逸舒适为主，随性而自在，身处其中能得到精神上的享受和放松。

业主希望能给孙女一个天真烂漫、充满梦幻与公主气息的房间，浪漫的欧式风格自然地流露在孙女房的每一个细节之上，整个房间的布置都恰到好处，精致舒适，以粉色为主基调，迎合了业主期望能培养出孙女的优雅气质的想法。

儿子房为现代风格，简约而大气，属于比较独特的现代家居风格，深受业主儿子的欣赏与青睐。

孙子房定制为现代轻奢风格，冷灰色调打造基底，呈现出大气的格调之余，意在给空间营造一种大大方方的利落感觉，风格也比较符合家人对男孩子的期待。

The master bedroom is filled with the natural atmosphere, and the design centers on ease and comfort, casual and unrestrained. Being in it, it is easy to get spiritual enjoyment and relaxation.

The owner hopes to give the granddaughter an innocent, dreamy and princess-like room. The romantic European style naturally reveals every detail of the granddaughter's room. The layout of the whole room is just to the point, exquisite and comfortable, with pink as the main tone. It caters to the owner's expectation to cultivate an elegant temperament of his granddaughter.

The son's room is a modern one, simple and majestic. It belongs to a relatively unique modern home style and is deeply appreciated and favored by the owner's son.

The grandson's room is a custom-made modern luxury style. The cool gray tone dominates the room and presents an grand style. It is intended to create a generous and agile perception of the space, and the style is more in line with the family's expectations for boys.

见与不见

Meet Me or Not

本案业主是一对年轻夫妇，希望房子设计得精致一些，生活要有仪式感。业主对设计师给予了极大的信任，全权交给他负责。客厅是生活中沟通的场所，一整面的储物加展示功能的柜子，在满足了生活储物需要的前提下，还拥有视觉美感。主卧室套间把阳面的小卧室做成了主卧的衣帽间，强调的是主卧室的舒适感。

项目名称：见与不见
项目设计：广东星艺装饰集团
项目地址：河北秦皇岛
设计师：刘天亮

The owner of this scenario is a young couple. They hope that the design of the house will be delicate to reveal the chase of ceremonial life. The owner has given great trust to the designer and delegated full responsibility to him. The living room is a place for communication. A cabinet with storage and display functions on a whole wall also has a visual beauty while meeting the needs of life storage. By making the small bedroom on the sunny side into the cloakroom of the master bedroom suite, the designer emphasizes a sense of comfort of the master bedroom.

郎成·永宁公馆

Lushine-Yongning Mansion

本案定位为年轻多元化的北欧风格。以白色为底色，用温暖的原木来中和白色的冰冷感，增加空间的暖色调。现代简约的明亮与北欧的人文工艺质感相结合，使空间显得干净整洁，没有多余的烦琐装饰，实现形式与功能的统一。餐厅的设计将北欧的朴实与工业风的复古相结合，创造出一种"另类"的美，座椅及餐桌的靓丽色彩让人食欲大增，诠释了一种个性张扬、自由不羁的新时代生活方式。

This scenario is positioned as a young and diverse Nordic style. White is used as the background color, and warm logs are utilized to neutralize the coldness of white and to increase the warm tone of the space. The combination of modern brightness and Nordic humanistic craftsmanship makes the space look clean and tidy and realizes the unity of form and function without unnecessary cumbersome decoration. The design of the restaurant combines the simplicity of Nordic style with the retro feeling of industrial style to create a "strange" beauty. The colorful seats and dining table stimulate people's appetite considerably, interpreting a new era of individualism and freedom lifestyle.

项目名称：邸成·永宁公馆
项目设计：广东星艺装饰集团
项目地址：福建龙岩
设计师：熊峰

山蓝

Mountain Blue

　　本案位于山边，从南面阳台及窗户望出去可以看到高大的灌木以及山景。室内以唯美独特的山蓝色调穿插，保证了整体的统一性，为整个空间注入一抹平和与宁静。开放的公共区域，户外光线随着落地窗进入室内，使室内的北欧气息得到烘托；空间中央的半高电视背景墙让视野更加开阔，使空间更有层次感，并让领域之间有更多互动，室内气息更显流动，谈心休憩更加自如惬意。

This scenario is located on the side of a mountain and tall bushes and mountains can be viewed from the balcony and windows on the south side. The interior is interspersed with beautiful and unique mountain blue to ensure the overall unity of color tone, instilling a touch of peace and tranquility into the entire space. In the open public area, the outdoor light enters the room through the floor-to-ceiling windows, which enhances the Nordic atmosphere of the interior. The half-height TV background wall in the center of the space widens the view, making the space appear broader, allowing more interaction between the fields and increasing the flow of the indoor air, thus to ensure a more relaxed talk and rest.

项目名称：山蓝
项目设计：广东星艺装饰集团
项目地址：广东广州
设计师：余林林

主卧采用山蓝色调背景墙，与深色床头柜、床、梳妆台、窗帘等互衬融合，营造室内华美厚重的气氛，给人一种舒适与惬意的感觉。

The master bedroom adopts a mountain blue background wall, which is integrated with the dark bedside tables, bed, dressing table, curtains, etc., creating a gorgeous atmosphere in the room, giving people a comfortable and cozy feeling.

保利天鹅语

Poly Swan's Words

本案空间面积约 450 平方米，分地下一层、地面两层。一层为别墅的核心，也是家的公共区域，结合古典简化的线条，配以家具打造出时尚、清新的法式混搭风格。负一层以娱乐为主，设红酒 BAR、茶室、架子鼓室、影视厅、工作间、阳光房等，采用大面积的奥特曼大理石作为墙面和地面材料，影视厅和茶室体现其功能的同时，采用防潮吸音布及防潮壁布，以避免潮气问题。二层为家庭休憩空间，整体考虑以温馨、舒适为主，打造一个静心减压的私人空间。

The space of this scenario is about 450 square meters, divided into one basement and two ground floors. The first floor is the core of the villa and also the public area of the home, combining classical and simplified lines and furniture to create a fashionable and fresh French mash-up style. The basement is mainly for entertainment, equipped with wine BAR, tea room, drum room, film and television hall, workshop, sunroom, etc. It uses a large area of ottoman marble as wall and floor materials. When film and television hall and tea room work, the moisture-proof sound-absorbing cloth and moisture-proof wall cloth are used to avoid moisture problems. The second floor is a family rest space, considered as a warm and comfortable area, creating a private space for calm and relaxing.

项目名称：保利天鹅语
项目设计：广东星艺装饰集团
项目地址：上海嘉定
设计师：黄剑

居心所·洪宅

Harbor-Hong's Residence

让生活回归本真。本真，意为本源，有纯洁真诚之意。当下，灯红酒绿的繁华世界里，或熙熙为名来，或攘攘为利去，众人只低头为生存而生活，约翰·列侬曾说："当我们正在为生活疲于奔命时生活已离我们而去。"

本案在墙面的装修中，用大量的木饰面装饰空间，使得整个空间被木质感包围，带来一种前所未有的温馨与朴素感。被阳光浸染的大露台，可供主人邀请三五知己，享受午后悠闲的下午茶，或者做一次瑜伽放松身体，来一场心灵上的净化。

Let life return to its authenticity nature. Authenticity means the origin, and it represents the meaning of pure and sincere. At the moment, in the bustling world of feasting, either for fame or for profit in the hustle and bustle of life, everyone bows their heads and lives for survival. John Lennon once said, "When we are struggling for life, life is gone. "

In the wall decoration of this scenario, a large number of wood veneers are used, making the entire space surrounded by wood, bringing an unprecedented sense of warmth and simplicity. The large terrace drenched by the sunshine can be used to enjoy a leisurely afternoon tea with several friends or serves as a place for yoga to relax the body and purify the soul.

项目名称：随心所·洪宅
项目设计：广东星艺装饰集团
项目地址：广东广州
设计师：丁捷

轻奢·2701

Mild Luxury-2701

项目名称：轻奢·2701
项目设计：广东星艺装饰集团
项目地址：四川内江
设计师：刘洋

业主是一对 80 后年轻夫妇，追求时尚。原始结构的卧室通道很窄，通过改造加宽形成一个二次玄关。将客厅原有的茶室改造成餐厅，餐厅和厨房之间的吧台满足了多功能的用餐需求，同时也便于传递和操作，整个空间显得更加宽敞。此外，客厅和餐厅地面用灰色大理石搭配玻璃和不锈钢，金属与石材给人一种强烈的视觉感召力；布艺沙发和硬包的运用均衡了空间里面材料的质地，同时也兼顾了女主人对家庭温暖的追求；橘黄色的单人沙发点缀活跃了空间的整体氛围，消除灰色和蓝色带来的低调冷感。空间中有刚柔、有冷暖，这应该是对轻奢最好的表达吧。

The owner is a young couple born in the 1980s chasing after fashion. The bedroom passage of the original structure is very narrow, and it has been transformed and widened to form a secondary entrance. The original tea room in the living room is transformed into a dining room. The bar table between the dining room and the kitchen meets the multi-functional dining needs, and is also easy to transfer the dishes, making the entire space more spacious. In addition, the floor of the living room and dining room uses gray marble to match with glass and stainless steel which gives people a strong visual appeal. The use of fabric sofas and background wall of hard materials balances the texture of the materials in the space, while also taking into account the hostess's pursuit of warmth. The orange single sofa embellishes the overall atmosphere of the space and eliminates the coldness brought by gray and blue. Hardness with softness and warmth with coldness in the space perfectly expresses the meaning of mild luxury.

贵州毕节兴隆家园

Xinglong House in Bijie, Guizhou

本案以新古典设计手法贯穿空间主线，用白色和米色搭配营造唯美浪漫、纯净高贵的空间美感。通过壁炉、家具、水晶灯、石材等细节，演绎精致与高贵，既延续古典风格又不拘泥于传统的思维逻辑，色调浅淡明快，机理清晰并富有质感，营造不一样的品位空间。

This neo-classical design technique is used throughout the design process, and the designer uses white and beige to create a beautiful, romantic, pure and noble space beauty. Every detail of fireplace, furniture, crystal lamp, stone, etc., reveals a refined and exquisite sense, which not only continues the classical style but also not bounded by the traditional thinking logic. With light and bright colors, clear and rich texture, a special and tasteful space is created.

项目名称：贵州毕节兴隆家园
项目设计：广东星艺装饰集团
项目地址：贵州毕节
设计师：冯立龙

纷华靡丽

Magnificent Residence

本案以浪漫主义为形式基础，材料多用大理石。多彩的织物、精美的地毯、精致的法国壁挂，烘托出空间豪华、富丽的视觉效果。用轻快纤细的曲线装饰，突出典雅、亲切的视觉感受。配饰上，金黄色和棕色的配饰衬托出古典家具的高贵与优雅，具有古典美感的窗帘和地毯、造型古朴的吊灯使整个空间看起来富含韵律感且大方典雅，柔和的浅色花艺为整个空间带来柔美的气质，给人以开放、宽容的非凡气度，让人丝毫不觉局促。

This scenario is based on romanticism, and the material is mostly marble. Colorful fabrics, artistic carpets, and exquisite French wall hangings create a luxurious and splendid visual effect. The whole space is decorated with light and slender curves, highlighting the elegant and cordial visual experience. As for accessories, golden and brown ones bring out the nobility and elegance of classical furniture. Curtains and carpets with classical aesthetics, and simple chandeliers make the whole space full of rhythm and elegance. The soft light-colored floral art reveals a mellifluous temperament, giving people feeling of open, tolerant and extraordinary rather than restrained.

项目名称：纷华靡丽
项目设计：广东星艺装饰集团
项目地址：广东广州
设计师：潘丽环

理想中的湾

Ideal Harbor

人们常说轻奢是"理性的品位"，它总是以优雅的面目呈现在我们面前，而美式带来的自由感则让这份"理性"更为温和、惬意、低调、有气质。

设计的形式有很多种，在整个设计中，以色块的选择作为手法，用米白色与蓝灰色作为对比，不管是客厅还是餐厅，均用这两种颜色不间断变换，以营造丰富的空间层次感。另外，大量的石膏线条，以及护墙板的线条和金属线条的叠加，让整个空间的层次感更强。加上香槟金的推拉门，以及后期搭配的家具作为呼应，让空间的颜色搭配更加协调。

People often say that mild luxury is a "rational taste" and it is always presented in front of us with an elegant appearance. Moreover, the sense of freedom brought by the American style makes this kind of "rationality" more gentle, cozy, low-key and refined.

There are many forms of design. In the whole design, the selection of color blocks is the unique technique. The beige and blue-gray are used as the contrast in the living room and the dining room. These two colors are continuously changed to create a rich layered space. In addition, a large number of plaster lines, as well as the superposition of the lines of the wall panels and the metal lines, make the whole space more spacious. Moreover, the champagne gold sliding doors and the chosen furniture echo each other, making the color matching of the space more coordinated.

项目名称：理想中的湾
项目设计：广东星艺装饰集团
项目地址：广西南宁
设计师：罗芳

观澜御湖世家

Guanlan Yuhu Shijia

本案致力于在简约的设计理念下将多种材质进行拼接融合，配合不同的照明手段，营造和谐统一的空间氛围。另外，设计师在空间线条的处理上，尽力实现顶面、墙面、地面的对齐与延续，打破原有空间的界限，使空间在稳定协调的基础上多一些变化，满足客户对空间整体效果不拘一格的要求。

This scenario is dedicated to splicing and blending various materials under the minimalist design concept, with different lighting methods to create a harmonious and unified space atmosphere. In addition, the designer tries his best to achieve the alignment and continuity of the top surface, wall surface, and ground in the treatment of space lines, breaking the boundaries of the original space, making more changes on the basis of original stability and coordination, and thus satisfying the owners needs for various patterns on the whole.

项目名称：观澜御湖世家
项目设计：广东星艺装饰集团
项目地址：安徽蚌埠
设计师：边诗琪

贵族

Nobility

法式家居常用洗白处理与华丽配色，洗白手法传达出法式乡村特有的内敛特质与风情，配色以白色、金色和深色的木色为主调。结构粗厚的木制家具，如圆形的鼓形边桌、大肚斗柜，搭配抢眼的古典细节镶饰，呈现皇室贵族般的奢华。秉持典型的法式风格搭配原则，餐桌和餐椅均为米白色，表面略带雕花，配合扶手和椅腿的弧形曲度，显得优雅矜贵，在白色的卷草纹窗帘、水晶吊灯、落地灯、瓶插百合花的搭配下，浪漫清新之感扑面而来。

Whitewashing and gorgeous color matching are commonly used in household of French style. The whitewashing technique conveys the introverted characteristics and exotic style unique to the French country. The color matching is mainly based on white, gold and dark wood. Thick-structured wooden furniture, such as round drum-shaped side tables and large-bellied chest of drawers, are matched with eye-catching classical details, presenting royal and noble luxury. Adhering to the typical matching principle of French style, the dining table and dining chairs are all beige, with a slightly carved surface, matching the curvature of the armrests and chair legs, which looks elegant and noble. The white curly curtains, crystal chandeliers, and floor lamps with the lilies in vase, the romantic and fresh feeling blows across the face.

项目名称：贵族
项目设计：广东星艺装饰集团
项目地址：四川达州
设计师：李焕

华发国宾壹号私宅

Private Residence in Huafa Guobin Yihao

圣·埃克苏佩里说过："设计师知道自己的作品在臻于完美时，并不在于无以复加，而在于无从删减。"设计师在设计这个空间时尽可能克制，以极少的表现装饰，利用空间、材质、光线创造出不同的生活体验。客厅整体地面铺着纹理地砖，米色沙发与整体色调相称。电视背景墙利用木材与墙面瓷砖拼接，凸显空间整体感又增加了空间活跃度，让客厅呈现出一种简约舒适的轻松氛围。

Saint-Exupéry once said, " A designer knows he has achieved perfection not when there is nothing left to add, but when there is nothing left to take away."
The design of the space is as restrained as possible with little external decoration. Different life experiences are created by making use of space, materials, and light. The whole floor of the living room is covered with textured floor tiles, and the beige sofa matches the overall tone. Wood is assembled with wall tiles to create the TV background wall, highlighting the overall sense of space and increasing the spatial liveness, presenting a simple, comfortable and relaxing atmosphere in the living room.

项目名称：华发国宾壹号私宅
项目设计：广东星艺装饰集团
项目地址：广西南宁
设计师：黄兴源、程文芳

领地海纳珑庭张宅

Zhang's Residence in Lingdi Haina Longting

传统中式家居风格历经上千年的演变，可谓传统工艺与匠心的体现，造型线条的运用较为复杂多变，陈设上往往雕琢各式繁复精美的图样。虽然华美，却不适合大多现代家庭。而新中式是一种传承，亦是一种创新。它扎根于中国传统文化的土壤之中，取其精华，舍其烦琐，以现代人的审美眼光来弘扬中国古典文化。

Traditional Chinese home furnishing style has undergone thousands of years of evolution. It can be described as the embodiment of traditional craftsmanship and ingenuity. The use of modeling lines is complex and versatile, and the furnishings are often carved with various complicated and exquisite patterns. Although gorgeous, it is not suitable for most modern families. The new Chinese style is a kind of inheritance and innovation. It is rooted in the soil of traditional Chinese culture, taking its essence, abandoning its cumbersomeness, and promoting Chinese classical culture with the aesthetic vision of modern people.

项目名称：邻地海纳珑庭张宅
项目设计：广东星艺装饰集团
项目地址：广东佛山
设计师：张晓财

本案以黑白灰为主要色调，这种现代中式家居风格表现出当代人对中国传统文化的理解。黑白灰和新中式相遇，描绘出一幅幅写意的水墨画。躺在舒服的沙发上，光是欣赏着仿佛有生命般的水墨背景和细纹雪花白薄板瓷砖，就十分舒适惬意。抱枕、窗帘等物件显露出的一抹绿意丰富了整个空间。微风拂动，薄纱间倾泻的光线如精灵，鲜活且灵动。任千年的故事在指间流淌，这是一种享受。

This scenario uses black, white and gray as the main color. This modern Chinese home style shows contemporary people's understanding of traditional Chinese culture. Black, white and gray meet the new Chinese style, depicting freehand ink paintings. Lying on a comfortable sofa, appreciating the living-like ink background and the fine-grained white snowflake sheet tiles, it is very pleasant. A touch of green revealed by pillows, curtains and other objects enriches the entire space. The breeze is blowing, and the light pouring from the tulle is like a spirit, fresh and agile. The story of thousands of years flows between fingers, which is definitely a kind of enjoyment.

　　阳台营造一步一景、一步一诗的画面，寸步之间流淌着假山流水的诗意之美，把中式文化诠释得淋漓尽致。

　　软装配饰巧妙地运用"典雅蓝"，点缀蓝调的雅致美丽。诱人的蓝调展现着深远的意蕴。将蓝色与新中式结合，旨在打造典雅高贵的美韵。它的搭配要点在于：保留木纹肌理的天然质感与自然属性，以黑白色调、简约造型的方式呈现，更为巧妙地突出主题及需求。

　　蓝白的色彩搭配完美勾勒出一幅青花瓷般的陶醉美景。

The balcony creates a picture of a scene and a poem step by step, and the poetic beauty of rockery and water lingers over the steps, which fully interprets Chinese culture.

Soft decoration cleverly uses "elegant blue" to embellish the elegance and beauty of blues. The seductive blues reveal far-reaching implications. The combination of blue and new Chinese style aims to create an elegant and noble beauty rhyme. The key points of its matching are: retaining the natural texture and natural attributes of the wood grain texture, presenting it in black and white tones and simple shapes, more cleverly highlighting themes and needs.

The blue and white color combination perfectly outlines a blue and white porcelain-like intoxicating beauty.

2 住宅·方案设计作品

Residence-Scenario Design Works

华润天合

Huarun Tianhe

本案是连排别墅，业主是一对年轻夫妇，孩子三周岁。在与设计师沟通过程中，业主强调空间要有空灵感，要有留白，不用过多地装饰。

This scenario is a townhouse and the owner is a young couple with a three-year old child. In the process of communicating with the designer, the owner emphasized that the space should leave room for inspiration and be free of too much decoration.

项目名称：华润天合
项目设计：广东星艺装饰集团
项目地址：广东广州
设计师：夏芳芳

　　在构建空间上，设计重点突出人与人、人与环境的互融性、亲密性、趣味性和秩序性。

　　本案的空间特点是下沉式花园设计，净高6米，在空间规划时，设计师主要考虑有二：一是小孩太小容易发生危险，应尽量采取保护措施；二是空间的利用问题，在界定空间功能时把负一层空间一分为二。负一层界定为公共空间，包括餐厅、工人洗衣房和户外花园；负二层界定为男主人工作室及影音室。

In the construction of space, the design focuses on the integration, intimacy, fun and order between people and the environment.

The space feature of this scenario is a sunken garden design with a net height of 6 meters. On planning the space, the designer mainly considered two aspects: one is that the house is dangerous to the little child and thus the protective measures should be taken; the other is the use of space, so the basement first floor is divided into two areas according to the relevant functions. The basement first floor is defined as the public space, including the restaurant, workers' laundry room and outdoor garden; the basement second floor is defined as the male master's studio and audio-visual room.

普罗旺斯浪漫

Romantic Provence

本案为法式风格，法式风格如同法国这个国家一样，浪漫又迷人，仿佛普罗旺斯的阳光洒满家里，多处金色调点缀营造出尊贵的宫廷的感觉。

This scenario is in French style. The French style is just like France, romantic and charming, as if the home is full of the sunshine of Provence, and many golden tones embellishment create the feeling of noble palace.

项目名称：普罗旺斯浪漫
项目设计：广东星艺装饰集团
项目地址：贵州贵阳
设计师：夏爱伟

香堤荣府别墅

Rong's Villa in Xiangdi

本案从传统徽派民居建筑中汲取灵感，经过精简和提炼，结合现代民宿风格，打造出休闲、舒适、简约、质朴的住宅空间。在木质元素的运用方面，传统民居粗犷的木柱与现代细腻的木饰面形成对比，并在形式上做了很多不同的尝试，表达出对传统与现代、质朴与精致之间融合的思考。

This scenario draws inspiration from the traditional Huizhou residential buildings, through streamlining and refining, combines modern homestay style, to create a leisure, comfortable, simple and pristine residential space. In the use of wooden elements, the rough wooden pillars of traditional houses contrast with the modern and delicate wooden finishes, and many different attempts have been made in designing form, expressing the thoughts about the fusion of tradition and modernity, simplicity and refinement.

项目名称：吾悦荣府别墅
项目设计：广东星艺装饰集团
项目地址：安徽蚌埠
设计师：边诗琪

延伸

Extension

设计师认为，不同的人对生活的理解不同，阳光、风、灯光、空间尺度、色彩、材质这一系列通过设计研发组合拥有太多的可能性，设计者需要根据居住者的生活特点去创造性地设计更贴合居住者习惯的环境。

对坐的沙发摆放形式强调的是沟通与交流，客厅应该是爱的传播空间，不应该仅仅是一个看电视的场所。

对于强调实用性的家居生活，在考虑美观的前提下首先要考虑的是储物空间。沙发背后墙面与餐厅墙面的装饰柜除了担当客厅的审美功能之外，也具有储藏功能。

The designer believes that different people have different understandings of life. Various combination of sunlight, wind, light, spatial scale, color, and material could have many possibilities through design and research and development. The designer thus needs to creatively design a living environment suitable for the residents according to their living style.

The opposite placement of the sofa emphasizes communication and conversation. The living room should be a space for the spread of love, not just a place to watch TV.

For home life that emphasizes practicality, storage space is the first thing to consider based on aesthetics. In addition to contributing to the aesthetic function of the living room, the decorative cabinets on the wall behind the sofa and the wall of the dining room also have storage functions.

项目名称：延伸
项目设计：广东星艺装饰集团
项目地址：河北秦皇岛
设计师：刘天宽

曦城花语二区

The 2nd District in Sunrise Metropolitan

　　为营造更好的整体布局，设计师对原始的户型结构进行了大刀阔斧的改造，将原本的越层空间连贯起来，做出大宅的效果，这也成为本案最大的亮点。

　　负一层将主客厅改造为挑空阳光房，用隔断连通书房空间的设计方式，营造了身居雅室、书韵馨香的氛围。通过空间外扩设置阳光房，为负一层充分引进自然光，使空间增添了一分磅礴气息。

　　黑白灰搭配原木色营造出温润自然的视觉感受，不同材质的大块面搭配和灯光细节丰富了空间的层次。

In order to create a better overall layout, the designer has drastically renovated the original apartment, linking the original cross-floor space to create a larger house, which has become the finest highlight of this scenario.

On the basement floor, the main living room is transformed into a sunroom with high ceilings and the design method of connecting the study space with partitions creates an atmosphere of living in an elegant room surrounded by the fragrance of books. Through the extension of the space, the sunroom is built to fully introduce natural light into the ground floor, which adds a majestic atmosphere to the space.

The combination of black, white and gray with the natural wood color creates a warm and natural visual experience. The decorations in big blocks of different materials and the lighting details enrich the hierarchy of the space.

项目名称：曦城花语二区
项目设计：广东星艺装饰集团
项目地址：河北秦皇岛
设计师：郭宇

多功能厅提供健身及休闲娱乐的功能。会客厅、书房、休息室一应俱全,贯通的大空间,极具享受体验,这是大宅应有的气派。

一楼客厅与庭院连通,庭院景观尽收眼底。主卧空间依据户型和主人生活习惯进行重新规划,充分体现了设计因空间而异,为人服务。

设计师以简约的美学手法构筑一个具有多重含义的空间,同时也在引导房屋主人找到自己喜欢的生活方式。

阳光房是这个下叠户型设计中的视觉焦点,充分体现了设计师对地下空间价值的深度挖掘。阳光自然倾泻在空间之中,使整个生活环境充满美感与温馨。

设计师对室内格局进行变动,营造了舒适的居住环境,而软装和细节的处理则体现出对精致生活的追求,从而使这个饱含现代时尚感的居所散发出蓬勃的生命气息。

居住空间是一个载体,承载着与这个家庭有关的每一个事件、每一份情感,设计师所做的,就是通过设计与情感的交汇,为空间制造更多可能。

The multifunctional hall provides fitness and recreational functions with access to the living room, study room, and lounge. These rooms are linked together for a exceptional experience of enjoyment, which is the manner preferred by a large mansion.

The living room on the first floor is connected to the courtyard, with a panoramic view of the courtyard. The master bedroom is re-planned according to the layout and the owner's living habits, which fully reflects that the design varies from space to space with a fundamental goal to serve people.

The designer uses simple aesthetics to construct a space with multiple implications, while at the same time guiding the owner to find his/her favorite lifestyle.

The sunroom is the visual focus in the design of this house with basement floor, which fully reflects the designer's in-depth exploration of the value of underground space. Sunlight pours into the space naturally, making the whole living environment full of beauty and warmth.

The designer has changed the interior pattern to create a comfortable living environment, while the soft furnishing and details reflect the pursuit of exquisite life, so that this modern and fashionable residence exudes a vigorous atmosphere of life.

The living space is a carrier that carries every event and every emotion related to this family. What the designer does is to create more possibilities for the space through the intersection of design and emotion.

海博熙泰

Haibo Xitai

业主是一位时尚的中年男士，在家居装修方面有自己独到的见解，倾向于现代简约风格。

客厅以浅棕色作为整体主色，白色布艺沙发与柔和的地毯、棕色线型沙发与大理石茶几相互映衬，使整体既有家的柔软又有现代工业风的时髦感。

业主工作之余常邀三五好友到家闲叙，大面积的玻璃酒柜便于来客挑选。开放式厨房与吧台方便业主在聚会时为朋友们亲自烹饪几道可口的下酒菜。

The owner is a stylish middle-aged man who has his own unique insights on home decoration and favors modern and simple style.

The living room uses light brown as the main color. The white fabric sofa and the soft carpet, the brown linear sofa and the marble coffee table set off each other, making the whole space reveal softness of home and the fashionable sense of modern industrial style.

The owner often invites several friends to come home to chat after work. The large glass wine cabinet facilitates choices. The open kitchen and bar are convenient for the owner to cook some delicious dishes for friends during the party.

项目名称：海博熙泰
项目设计：广东星艺装饰集团
项目地址：广东潮州
设计师：徐源

　　薄荷绿与玫瑰粉相互碰撞，金色的包边线条与拼接地毯，右侧窗台的飘窗经久耐看，左侧的书柜方便学习，助力孩子的成长。衣帽间兼顾私人会客厅与休闲影音室的功能，成为放空、嬉笑的小天地。

Mint green and rose pink collide with each other. The golden edging lines match the spliced carpet. The bay window on the right side is durable and good looking and the bookcase on the left side is convenient for learning which could contribute to the growth of the child. The cloakroom takes effect on functions of private living room and studio room, making it a small place for emptying the mind and having fun.

中铁生态城白晶谷

China Railway Eco-City White Crystal Valley

本案以"韵"为主题，在整体空间和谐基调上彰显出浪漫的气质。软装与硬装上中式元素的运用，成就了本案的精髓与灵魂。设计力图在奢华优雅中呈现出中式的文化气息，追求深沉里显露尊贵，典雅中浸透奢华的设计表现。

项目名称：中铁生态城白晶谷
项目设计：广东星艺装饰集团
项目地址：贵州贵阳
设计师：金嘉宜

The theme of this scenario is "rhyme", showing a romantic temperament based on the harmonious tone of the overall space. The use of Chinese elements in soft and hard decoration is the essence and soul of this case. The design strives to present the Chinese cultural atmosphere with luxury and elegant decoration, pursuing the nobleness in depth and reflecting the luxury design with elegance.

盒子

Square Box

选择一种设计风格就是选择一种生活方式。干净整洁的空间，会让人的心态变得平和。设计师将功能美学化、艺术化，使生活除了柴米油盐酱醋茶之外，还增添了美学涵养，利用材质、光影使空间更显层次感。搭配家具与艺术品，让业主在一个低压、舒适的空间里获得精神的享受。设计师对亲子空间展开设计，使这个空间看起来就像是一个用积木搭起来的方盒子，又像是漂浮于空间中的小建筑，让原本单调的空间有了全新的体验。

A design style is a lifestyle. A clean and tidy space will make people feel peaceful. The designer makes the function aesthetic and artistic, so that in addition to the basic needs of firewood, rice, oil, salt, sauce, vinegar and tea, aesthetic appreciation is also added, using materials, light and shadow to make the space more layered. With carefully chosen furniture and artwork, the owner can get spiritual enjoyment in a low pressure and comfortable space. The designer has designed a parent-child space, making it look like a square box built with blocks as well as a small building floating in the space, enabling the originally monotonous space filled with a new experience.

项目名称：盒子
项目设计：广东星艺装饰集团
项目地址：广东广州
设计师：周燕

阅山湖别墅

Yueshanhu Villa

休憩空间设计基于对使用者的关照，让居住者融入空间故事中，与空间产生强烈的情感联结，唤起人们对愉悦生活的热爱与向往。

本案强调空间的整体性和风格的统一性。根据不同功能对相应的房间立面作出处理，提倡自然简洁和理性的规则，比例均匀，形式新颖，材料搭配合理，收口方式干净利落，维护方便。整个内部结构严密紧凑，空间穿插有序，围护体各界面要素的虚实构成比较明显，通过空间的虚实互换，取得局部与整个空间的和谐，强调空间的完整性和高贵典雅感。

The design of the rest space is based on the care for the users, allowing the residents to create a strong emotional connection with the space, and arousing people's love and yearning for a pleasant life.
This scenario emphasizes the integrity of space and the unity of style. Corresponding room facades are processed according to different functions, reflecting natural simplicity and rational rules with uniform proportions, novel forms, reasonable material collocation, smooth closing methods, and convenient maintenance. The entire internal structure is tight and compact, and the space is interspersed in an orderly manner. The virtual and real composition of each interface element of the enclosure is relatively obvious. Through the transformation of virtual and real space, harmony between the part and the entire space is achieved, emphasizing the integrity of the space and the sense of noble elegance.

项目名称：阅山湖别墅
项目设计：广东星艺装饰集团
项目地址：河北秦皇岛
设计师：刘春然

　　色彩上，主体色与点缀色和谐统一，艺术、大气而有品质。关注照明设计，使灯光不仅满足照明要求，更营造出一个富有艺术感的生活氛围，符合整体风格。

　　本案设置了新风系统、地热系统、中央吸尘系统以及声光等智能家居系统，体现出时代性、生态性、环保性和智能性，使科技节能与艺术品质完美结合。

In terms of color, the main color and the embellishment color are harmonious and unified, which is artistic, grand and high-quality. Lighting design is given special attention, so that the light not only meets the lighting requirements, but also creates an artistic living atmosphere that conforms to the overall style.

This scenario is equipped with a fresh air system, a geothermal system, a central dust collection system, and an intelligent housing system such as sound and light, which reflects the epochal, ecological, environmental and intelligent characters, and thus makes the perfect combination of energy saving and artistic quality.

花语

Language of Flowers

本案位于意桥岛江边别墅，主要建筑结构运用钢结构与钢化玻璃，使用轻质材质来达到透明可视的效果。别墅花园是男主人精心为太太打造的世外桃源，女主人比较喜欢花花草草，所以室内设计偏向简约，同时善用透明材质设计，这样主人就能随时感知到室外的景观变化。

This scenario is located in a villa on the riverside of Yiqiao Island. The main building uses steel structure and tempered glass and uses lightweight materials to achieve a transparent and visible effect. The villa garden is a paradise created by the host for his wife. The hostess is fond of flowers and plants, so the interior design tends to be simple and transparent material is used so that the owners can perceive the changes of outdoor landscape at any time.

项目名称：花语
项目设计：广东星艺装饰集团
项目地址：广东广州
设计师：郑杰勇

珠江花园·眺望

Pearl River Garden-Overlooking

本案临近珠江，在外观设计上打破传统思维，增加玻璃连桥和眺望台，使视觉更加开阔、造型更加突出，江景一览无遗。

室内空间布局以社交的概念呈现，讲究空间与空间的互动、环绕，室内与室外的开放性、一步一景是本案的设计重点。

This scenario is close to the Pearl River, breaking the traditional thinking in the exterior design, adding glass bridges and gazebos to widen the view as well as create unique shape with access to the unobstructed riverscape.
The indoor space layout is presented in the concept of social interaction, paying attention to the interaction and connection of space. The openness of indoor and outdoor space and the change of scene by step are the design focus of this scenario.

项目名称：珠江花园·眺望
项目设计：广东星艺装饰集团
项目地址：广东广州
设计师：潘丽环、陈热情

"DU 品"·凡尔赛

"DU Pin"-Versailles

设计师所理解的法式风格，其实就是保守加经典。在经典的基础上加入一些性感元素，就会有更多不一样的味道。本案采用法式风格设计，在色彩上做了大胆的冲突对比，独特的橙红色玄关、造旧白的客厅，色泽干净，鲜丽夺目。软装搭配上，简化了设计成分，让传统法式带有一些现代精简的视觉设计效果，使整体感觉更前卫。

The French style that the designer understands is the conservative plus classic style. Some sexy elements are added on the basis of classics, and there will be more different flavors. This scenario is designed in French style, with bold conflict and contrast in color. The unique orange-red porch and the old white living room are clean and dazzling. In the soft outfit, the design components are simplified, adding some modern and streamlined visual design effects to the traditional French style, making the overall feel of the space more avant-garde.

项目名称："DU 品"·凡尔赛
项目设计：广东星艺装饰集团
项目地址：重庆万州
设计师：邹珊

侨建·御溪谷

Qiaojian-Yuxigu

新中式是中国传统文化在当前时代背景下的演绎，是在对中国文化充分理解基础上的当代设计。新中式是传承传统中式风格的精髓，通过与现代潮流的对话碰撞而产生的创新设计。本案在设计上延续了两宋时期家居配饰理念，提炼了经典元素并加以简化和丰富，在家具选择上偏重于简洁清秀，同时打破了传统中式空间布局中等级、尊卑等文化思想。空间配色上使用了局部的红色点缀，凸显中国"魂"。

New Chinese style is an interpretation of traditional Chinese culture in the context of the current era, which is a contemporary design style based on a full understanding of Chinese culture. It is an innovative design style that inherits the quintessence of traditional Chinese style through exchange and collision with modern trends. The design of this scenario continues the idea of home accessories during the Song Dynasty and refines classic elements with additional simplification and enrichment. The furniture selection focuses on simplicity and elegance, while breaking the traditional Chinese spatial layout of hierarchy, senior and junior cultural ideas. Partial red embellishments are used in the color matching of the space, highlighting the Chinese "soul".

项目名称：协建·御溪谷
项目设计：广东星艺装饰集团
项目地址：广东广州
设计师：胡哲、侯立原

茗品

Ming Pin

本案线条简单清晰，巧妙地运用灯光的效果，使各式陈列品呈现非常独特的个性美和时尚美。在家具配置上，白亮光系列家具的独特光泽使家具时尚感倍增，带来舒适与美观并存的享受。在配饰上，延续了黑白灰的主色调，以简洁的造型、完美的细节营造出时尚前卫的感觉。线条有的柔美雅致，有的遒劲而富于节奏感，整个立体形式都与有条不紊的、有节奏的曲线融为一体。以宁缺毋滥的精髓，合理地简化居室，从简单舒适中体现生活的精致。而厅堂的设计则以多样化设计为主体，配以细节、灯光、色彩等元素，使整体家居风格更加异彩纷呈。

The lines of this scenario are simple and clear, and the effect of lighting is used cleverly to make all kinds of displays present a unique personality and fashionable beauty. In terms of furniture selection, the white light series, whose unique luster doubles the fashion sense of the furniture and brings the coexistence of comfort and beauty. As for the accessories, black, white and gray dominate the space, with the simple shape and perfect details to create a fashionable and avant-garde feeling. Some of the lines are soft and elegant, some strong and full of rhythm. The whole three-dimensional form is integrated with the orderly and rhythmic curves. Based on the idea of quality before quantity, the living room is rationally simplified revealing the exquisite life from simplification and comfort. The design of the hall is based on diversified techniques, with details, lighting, colors and other elements to make the overall home style more colorful.

项目名称：茗品
项目设计：广东星艺装饰集团
项目地址：四川成都
设计师：谢志兵

入户门厅采用红色的大理石装饰墙面，增加视觉冲击力的同时也让每个进入此空间的人有不一样的体验。硬装偏向现代风格，家具和软装则偏向古典风格，整体给人一种时尚、奢华、有品位的视觉感受。

现代轻奢虽也是现代风格的一种，但与传统的现代风格不同的是，现代轻奢在色彩运用上对比鲜明，常用高级黑与高级灰作为底色，给空间营造出一种利落的感觉，加上明度比较高的驼色、象牙白等，为居室呈现出一种温馨大气、时尚高雅的氛围。现代轻奢风格比较注重生活的质量，装修上一般简洁奢华，为每一位居住者打造一个身心放松的家居环境。

茶台背后是瓦片结合大理石设计成的茶具柜。精致的挂画、古典的装饰品、镂空的屏风，搭配采用中式布局灯光设计的天花，全面透露出中式古典气息。

The entrance hall uses red marble to decorate the wall, which increases the visual impact and allows everyone who enters the space to have a different experience. Hard furnishings tend to be modern, while furniture and soft furnishings tend to be classical, giving people a stylish, luxurious and tasteful visual experience.

Although modern mild luxury is also a kind of modern style, it is different from the traditional one in that modern luxury has a sharp contrast in the use of colors. High-grade black and high-grade gray are often used as background colors to create a neat feeling in the space. Moreover, the use of camel and ivory with high brightness, presents a warm, stylish and elegant atmosphere. The modern mild luxury style pays more attention to the quality of life, and the decoration is generally simple and luxurious, creating a relaxing home environment for every resident.

Behind the tea table is a tea set cabinet designed with tiles and marble. Exquisite hanging paintings, classical decorations, hollow screens, and ceilings with Chinese-style layout lighting design, fully reveal the classic Chinese flavor.

　　中国风的构成主要体现在传统家具（多以明清家具为主）、装饰品及以黑色和红色为主的装饰色彩上。室内多采用对称式的布局方式，格调高雅，造型简朴优美，色彩浓重而成熟。中国传统室内陈设包括字画、匾幅、挂屏、盆景、瓷器、古玩、屏风、博古架等，体现一种修身养性的生活境界。中国传统室内装饰艺术的特点是总体布局对称均衡、端正稳健，而在装饰细节上崇尚自然情趣，花鸟、鱼虫等精雕细琢，富于变化，充分体现出中国传统美学精神。设计师在设计时在传统生活、文化习惯和精神意识层面进行思考，关注传统精华空间的继承，试图用现代主义的句法将古典语汇与现代语汇相结合，提升空间品质。

Chinese style is mainly reflected in traditional furniture (mostly Ming and Qing furniture), decorations and black and red colors. The interior space mostly adopts symmetrical layout, with elegant style, simple and beautiful shape, strong and mature color. Traditional Chinese indoor furnishings include calligraphy and painting, plaques, hanging screens, bonsai, porcelain, antiques, screens, and ancient frames, etc., reflecting a life state of self-cultivation. Chinese traditional interior decoration art is characterized by symmetrical and balanced as well as upright and stable layout and advocating natural taste in decorative details, including flowers, birds, fish and insects, which are meticulously crafted and full of changes, fully embodying the spirit of traditional Chinese aesthetics. When designing, the designer takes account of traditional life, cultural habits and spiritual consciousness. Meanwhile, he pays special attention to the inheritance of traditional essence and tries to use the syntax of modernism to combine classical vocabulary with modern vocabulary to improve the quality of the space.

A10-3102

A10-3102

本案设计展现出极简主义的稳重和不失惊艳的韵味。通过木质元素的充分运用、高级灰调的质感以及扑面而来的轻奢优雅，质朴自然的现代设计融合当代美学理念，营造出一种轻奢的生活氛围。

The design of this scenario shows the stability of minimalism and the stunning charm. Through the full use of wooden elements, the texture of high-grade gray tones and the light and luxurious elegance, the simple and natural modern design combines contemporary aesthetics to create a mild luxury living atmosphere.

项目名称：A10-3102
项目设计：广东星艺装饰集团
项目地址：广西南宁
设计师：周平枰

水映长岛别墅设计

Villa Design in Shuiying Changdao

设计别墅时，如果硬要往里堆砌豪华的建材，用星级酒店的商用标准来设计，不仅耗费巨资，还会让人产生不适感。家和酒店毕竟是有区别的，家是体现温馨舒适的场所，是每一个角落都能让居住者放松的地方，因此，适合居住应该放在首位。

另外，设计师表示，舒适空间应具备六个特征：一是功能空间要实用；二是心理空间要实际；三是休闲空间要宽松自然；四是自然空间要陶冶精神，放松心情；五是生活空间要以人为本；六是锁定私密空间，满足人性最大限度的空间释放。

When designing a villa, if one insists on piling up luxurious building materials and tries to design it with the commercial standards of a star hotel, it will not only cost a lot of money, but also make people feel uncomfortable. After all, there is a difference between a home and a hotel. Home is a warm and comfortable place. It is a place where residents can relax in every corner. Therefore, suitable for living should be the highest priority.

In addition, the designer stated that a comfortable space should have six characteristics: first, the functional space should be practical; second, the psychological space should be realistic; third, the leisure space should be relaxing and natural; fourth, natural space should cultivate spirit; fifth, the living space should be people-oriented; sixth, a private space is needed to satisfy the human natural need of the maximum freedom.

项目名称：水映长岛别墅设计
项目设计：广东星艺装饰集团
项目地址：四川成都
设计师：谢洪

双河湾

Shuanghe Bay

本案以高级灰为主调，搭配精致的家具、利落的线条，以及木材、金属、玻璃或人造材料，凸显出空间的时尚简约范儿。没有花哨的造型，也没有浓重的色彩，单是简单而不失质感的空间设计，就给人带来一种宁静高贵的视觉感受。整个设计风格摒弃了烦琐浮华，以一种简约优雅的设计语言，结合对材质的精彩运用，极力呈现出低调、奢华的空间力量。

优雅是一种放弃，放弃抢眼的色彩，放弃夸张的造型，放弃出风头，放弃自夸，放弃炫耀，放弃刻意表达，让随意与优雅成为习惯。家凝聚着万千诗意，它美得不可方物，它是艺术也是生活。所谓奢华，不在雕墙峻宇，而在于空间所传递的品质与态度。

客厅、餐厅用多幅水墨画作装饰，丰富了空间画面感，素雅的色彩与整体色调互相呼应，显得更加典雅大气。设计师基于高级灰色调的整体空间氛围，在家具选择上，加入了低彩度的橙色作为点缀，给室内注入活力。大面积的白墙与高级灰形成碰撞，带给人极致奢华的空间体验。

This scenario is based on using high-grade gray, with exquisite furniture, clean lines, and wood, metal, glass or man-made materials, highlighting the fashion and simplicity of the space. There are no fancy shapes and no strong colors. The simple and textured space design gives people a peaceful and noble visual experience. The whole design style abandons the cumbersome and vanity, and uses a simple and elegant design language, combined with the wonderful use of materials, to show a low-key and luxurious space power.

Elegance is a form of giving up, giving up eye-catching colors, giving up exaggerated styles, giving up the limelight, giving up boasting, giving up showing off, giving up deliberate expression, and letting casualness and elegance become a habit. The home is full of poetry. It is so beautiful and it is not only an art but also life. The so-called luxury does not lie in the majestic carved walls, but in the quality and attitude conveyed by the space. The living room and dining room are decorated with multiple ink paintings, which enriches the sense of space. The elegant colors and the overall tone echo each other, making it more elegant and grand. Based on the overall space atmosphere of high-grade gray tones, the designer has selected low-color orange furniture as an embellishment to inject vitality into the interior. Large areas of white walls collide with high-grade ash, giving people an extremely luxurious space experience.

项目名称：双河湾
项目设计：广东星艺装饰集团
项目地址：云南昆明
设计师：熊卫星

葡萄岛小镇样板间

Sample Room of Grape Island Town

本案将居住者对未来生活的美好想象凝聚于窗外的一片自然海景上，自在的动线、自然的感受，抱海天景致，给人以开放透气之感。打造"高品质度假生活"是设计师对本案最初始的想法。

设计师在原有结构框架基础上进行了充分的利用与改造，重新整理墙面及规划格局，模糊了室内外界线，将大量采光及室外景色带进屋内。以一种持续的方式，重新创造了"贯通"于建筑内的全新空间体验，使居住者在住宅中穿行的同时与场地的极佳位置产生紧密联系，于偶然间邂逅周边天然海景，于质朴纯净中体味生活。

The natural seascape outside the window satisfy the residents' imagination of beautiful future life in this scenario. The fine room and natural feeling, with the view of sea and sky, giving people a sense of openness and breathability. Creating a "high-quality vacation life" is the designer's initial idea for this scenario.

The designer has made full use and renovation of the original structural framework. The walls and the planning layout have been rearranged, which has blurred the indoor and outdoor lines, and make access to a lot of daylight and outdoor scenery into the house. In a continuous way, a brand-new spatial experience "permeated" in the building is recreated. Therefore, when the residents walk through the house, they could have a close connection with the excellent location of the site, encountering the surrounding natural seascape by chance as well as experiencing life in simplicity and purity.

项目名称：葡萄岛小镇样板间
项目设计：广东星艺装饰集团
项目地址：河北秦皇岛
设计师：崔健、夏文彬

　　空间中没有太过抢眼的设计，静静地衬托与融入自然海景的美，造就疗愈人心的自然居所。材料运用上充分考虑空气湿度等主要因素，以天然抗潮湿装饰材料为主，原木、米黄、白色，简洁的配色组合，素雅清爽，营造出温馨安逸的空间氛围，在木色的舒缓勾勒中，愈发呈现自然亲和的气质。天然通透材质的隔断界定着起居和休憩的空间划分，让起居、就餐、社交、休闲等机能充满弹性及变化，尽可能减少表象装饰，利用空间、材质、光线创造出不同于寻常的干净生活感受，或者说是一种净化过的内在哲学体验。

There is no eye-catching design in the space. The natural seascape is quietly integrated into the whole design, creating a healing natural residence. In the use of materials, the main factors such as air humidity are fully considered and the natural anti-humidity decorative materials are mainly used. The simple color combination of logs, beige and white, is elegant and refreshing, creating a warm and comfortable space atmosphere. With the soothing outline of wood color, the space sends out a natural and friendly temperament. The partition of natural transparent material defines the space division between living and resting, making living, dining, social, leisure and other functional areas full of flexibility and change. The design has minimized exterior decoration and used space, materials and light to create an unusual neat life feeling or a purified inner philosophical experience.

心居

Soul Harbor

本案通过明朗简约的线条，将空间进行了合理的划分。面对扰攘的都市生活，有一处能让心灵沉淀的空间，是业主心中的一份渴望，也是本设计在方案中所体现的主要思想。因此开放式的大厅设计给人以通透之感，避免视线给人带来的压迫感。没有夸张，不显浮华，通过干干净净的设计手法，将业主的工作空间融入生活空间中。

In this scenario, the space is reasonably divided by clear and simple lines. Facing the hustle and bustle of urban life, a space where the soul can settle is a desire of the owner and the main idea embodied in this design. Therefore, the open hall gives people a sense of transparency, avoiding the oppressive feeling of sight. There is no exaggeration, no flashiness, and the owner's working space is integrated into the living space through neat design techniques.

项目名称：心居
项目设计：广东星艺装饰集团
项目地址：内蒙古包头
设计师：梁俊

色阶

Tone Scale

本案以黑白灰的极致三色调为主题，紧跟当代时尚的步伐。"白"的简洁明快，"黑"的精致优雅，"灰"的自然柔和，三色调在空间中相辅相成，凸显当代家居设计去繁从简的态度。四季皆宜的黑白灰装饰褪去缤纷色彩，将空间化为一种低调的奢华。

平静不失深刻，家居历久弥新，彰显居住者生活的精致与个性。材质的选择与应用追求道法自然的哲学思维，以现代工艺材质引入自然的气息，让现代空间与自然紧密结合，使人的生活得到"质"的突破，实现放松身心、愉悦自我的目的。

The theme of this scenario is three ultimate tones of black, white and gray, keeping pace with contemporary fashion. "White" is concise and lively, "black" is exquisite and elegant, and "gray" is natural and soft. The three tones complement each other in the space, highlighting the contemporary home design's attitude of going after simplicity. Without glorious colors, the black, white and gray decorations are suitable for all seasons, turning the space into a low-key luxury area.

Home furnishings are timeless, calm and profound, showing the exquisiteness and individuality of the residents' lives. The selection and application of materials pursues the philosophy of Taoism and introduces natural elements with modern craft materials, so that the modern space and nature are closely integrated. Moreover, people's life can get a "qualitative" breakthrough, and realize the purpose of relaxing the body and mind and enjoying oneself.

项目名称：色阶
项目设计：广东星艺装饰集团
项目地址：广东广州
设计师：魏己成、骆舜尧

都会新贵系

Urban Elite Series

解读时代语言，在"房住不炒"的背景下，楼市趋冷，消费回归理性，使高端楼市产品转向居住生活定制。社会演进决定生活方式的更迭，当代中国的大都会背景下，都会新贵快速崛起，其生活方式亦应运而生。

本案保留古典传统的造型与线条，简洁利落地赋予空间以时尚面貌，更重要的是它依然带有奢华风骨，是在传统与现代间展现都会新贵的上佳选择。把艺术生活化，通过高级的设计手法表达出来。美好的生活关系，引导生活理念，跟随时代，追随内心。不在意外物，而关心与家人的关系。

Under the background of "housing to live without speculation", the property market is getting cooler in the new era. The consumption returns to rationality and high-end property products are shifted to residential life customization. Social evolution determines the change of lifestyles. Under the metropolitan background of contemporary China, urban elites are rising rapidly and thus their lifestyles also emerges.

This scenario retains the classical and traditional shapes and lines. A space of a fashionable look is revealed through these neat and concise lines. More importantly, it still has a luxurious core. It is a good choice for showing the elites between tradition and modernity. Turn art into life and express it through advanced design techniques. A good life relationship guides life concepts, follows the times and follows the heart. The relationship with family members is always the highest priority without any interruption of external world.

项目名称：都会新贵系
项目设计：广东星艺装饰集团
项目地址：山西大同
设计师：艾仁

万科朗润园别墅

Villa in Vanke Spring Dew Mansion

本案对空间功能性、情感性、观赏性颇为注重。以业主生活习性切入，拆除原有非承重墙体，对各空间进行动静划分和布局整合。设计师以归零思维，发掘空间最大可能性，并适时衍生惊喜感。在结构改造中，本设计追求阳光、空气、风景能自由畅快流动，注入层次感和自然感。摆设从简，在平静且克制的基调中，创造空间融合及视觉张力。

This scenario pays much attention to spatial functionality, emotion and appreciation. According to the living habits of the owner, the original non-load-bearing walls were removed, and the dynamic and static divisions and layout integration of the spaces were carried out. The designer takes a clean-slate approach to explore the greatest possibilities of the space and creates a sense of surprise from time to time. In the structural transformation, the design centers on the natural flow of sunlight, air, and scenery, injecting a sense of hierarchy and spontaneity. The furnishings are simple, creating spatial integration and visual tension in a calm and restrained tone.

项目名称：万科朗润园别墅
项目设计：广东星艺装饰集团
项目地址：广东中山
设计师：杨琛

　　出于趣味性考虑，采用天井的方式连通一层与二层。让进门口的位置更富有仪式感，带来精神世界的营养供给。偌大的墙身使用大板砖进行铺贴，使得空间更加开阔大气。使用深色面板的拼接，让色彩之间产生强烈的碰撞，带来视觉上的冲击。

　　平静的色块给予空间更多舒适和放松，让人乐享惬意时光。在这样充分调动个人需求与情感的家中，所有设计考究处都将带来会心一击，空间是有生命的，踏入的一瞬，也就在与之对话了。

Considering interestingness, the first and second floors are connected by a patio. The entrance is made more ritual, which can inject energy to the spiritual world. The huge walls are paved with large slabs to make the space more spacious. The splicing of dark panels creates a strong collision among colors and brings about a visual impact.

Calm color blocks endow the space with more comfort and relaxation, enabling people to enjoy a pleasant time. In such a home that fully mobilizes personal needs and emotions, all the well-designed parts will bring a critical hit. The space is lively, and the moment you step into it, you will be in communication with it.

轻奢是一种态度

Mild Luxury, an Attitude

本案业主是一位年轻时尚的职业女性，有主见，有个性。经过几次设计理念的沟通，最终与设计师敲定了现代轻奢风格。对于精英阶层而言，豪奢不是最重要的，传达一种充满"精英感"的生活态度或许更为重要。因此比起视觉表面的迎合，凸显品位与个性，拥有适宜的大小、丰富的功能、舒适的氛围才能打动他们。

项目名称：轻奢是一种态度
项目设计：广东星艺装饰集团
项目地址：海南海口
设计师：邓佑

The owner of this scenario is a young and fashionable professional woman who is strong-minded and individualistic. After several communication of the design idea, the modern mild luxury style is confirmed. For the elite, luxury is not the most important thing. It may be even more important to convey a life attitude full of "sense of elite". Therefore, instead of catering to the superficial visual impact, the designer must highlight the taste and personality with appropriate size, rich functions, and comfortable atmosphere to impress them.

客餐厅设计具有舒适的功能性，拥有超强储物功能的电视背景墙增强了客厅活跃性；简约而不简单的沙发背景使整个客厅主次分明，吊顶金色铜条收边与墙面、地面形成了完美的融合，墙面主色调以粉色系为主，保留了业主的一颗少女心。

在卧室的设计上，强调的是舒展与放松，没有刻意的豪华，但流露出精致的生活情调。

主卧套房内设衣帽间和主卫，各功能区呈递进式布局。在氛围上，营造时尚而不浮躁、典雅却不乏浪漫的感觉。

卫生间区域采用了干湿分离的功能布局。以白色大理石上墙，配上铜色五金件，整体格调简单中透露着奢华。

设计师采用现代的设计手法，衍生出一个具备文化性、象征性、装饰性的空间。

The design of the guest dining room is comfortable and functional. The TV background wall with super storage function enhances the living room's vitality. The concise but not simple sofa background makes the whole living room distinct. The golden copper strips on the ceiling integrate with wall and ground perfectly. The main color of the wall is mainly pink, which retains the owner's girlish heart.

In the design of the bedroom, the emphasis is on stretching and relaxation, without deliberate luxury, but revealing exquisite lifestyle.

The master bedroom suite has a cloakroom and a main bathroom, and each functional area has a progressive layout. It creates a stylish but not impetuous, elegant and romantic feeling.

The toilet area adopts a functional layout of separation of dry and wet. With white marble on the wall and copper-colored hardware, the overall style is simple and reveals luxury.

The designer uses modern design techniques to derive a cultural, symbolic and decorative space.

3 公共·工程实景作品

Public-Engineering Live-Scene Works

一大街亚朵酒店

Atour Hotel on the First Avenue

一大街亚朵酒店共9层，有104间客房。亚朵是一家令人舒心的微笑酒店，提倡人文、温暖、有趣的"在路上"第四空间生活方式，致力于向新中产消费者提供优质的酒店服务和生活方式产品。

Atour Hotel on the First Avenue has 9 floors and 104 rooms. Atour is an agreeable hotel that advocates a humanistic, warm, and interesting "on the way" lifestyle. It is committed to providing high-quality hotel services and lifestyle products to new middle-class consumers.

项目名称：一大街亚朵酒店
项目设计：广东星艺装饰集团
项目地址：天津
设计师：陈明

聚商·公园壹号

Jushang-First Park

本案坐落在贵州一座离贵阳车程40分钟的县城内，该县城是一个少数民族聚集之地，当地气候湿润、怡人，光照时间长，整体温差不大。整个小区的位置空旷平缓，十分适合进行大型商住综合体的开发，售楼部建筑物呈狭长形，就现场的尺寸比例来说调整格局难度并不大。甲方要求拿出设计改变雷同的结构，并增加能开展销售活动的大空间。经过实地考察后，设计师发现竞争楼盘的整体风格包括建筑物风格为新中式，在与甲方沟通和看过裙楼外形后，决定采用现代轻奢的风格作为本案的主体设计方向。

This scenario is located in a county town, about 40 minutes away from Guiyang by car in Guizhou. The county town is a place where ethnic minorities gather. The local climate is humid and pleasant, with long sunshine hours and little overall temperature difference. The location of the entire community is open and gentle, which is very suitable for the development of large-scale commercial and residential mall. The sales buildings are long and narrow, and the adjustment of the layout is not difficult in terms of the size. Party A requested that the design should change the similar structure and increase a larger space for sales activities. After on-site inspection, the designer found that the overall style of the competitive real estate, including the building style, was a new Chinese style. After communicating with Party A and seeing the shape of the podium building, he decided to adopt a modern mild luxury style as the main design direction of the scenario.

项目名称：聚商·公园壹号
项目设计：广东星艺装饰集团
项目地址：贵州贵阳
设计师：王硕

整个售楼部大厅采用极简的金属线条和云灰大理石为主体，用内嵌式金属踢脚线暗藏灯带。通过石材衬托空间的大气和开敞度，用极少的设计语言为后期的软装设计留下一个相对稳定的基层。木质护墙板采用当地的实木材料作为情感的寄托，让商业的氛围相对弱化。超大的采光面给大厅留出与自然的交流空间。本案结构上的设计安排为本次设计工作的中心，挑空大堂接近 20 米长，设计师难以对整个大厅进行更好的功能分区。在结构上，设计师大胆地进行了独立空间和架空层的结构理念设计，一方面对销售对象进行合理有序的分流，始终让大厅保证一定的奢华性；另一方面能保证销售的有序和洽谈的私密性。向层高要空间，让功能分区更有余地，也从结构上彻底打破了甲方竞争对手的单一——层式展示空间，并且让销售和样板房与办公层有一条相对流畅的动线，也能展示一个大房企的整体管理能力，为后期销售加分。色调上以灰色系为主，因为实地考察后发现甲方商业裙楼的大型整体外观是玻璃幕墙加隐形 LED 屏幕，考虑到当地的地理位置、采光以及温度等因素，设计师决定让整个空间表达的色彩语言给人一种硬朗和偏冷的感受，这样既保证了降低暖通设备投入的甲方需求，也能最大化地节能。

The entire lobby of the sales department uses minimalist metal lines and grey marble as the main body. The design uses built-in metal skirting to hide the light belt. The stone sets off the atmosphere and openness of the space, leaving a relatively stable base for the later soft decoration with very little design. The wooden wall panels use local solid wood materials as emotional sustenance, which makes the commercial atmosphere relatively weak. The large daylighting surface in the hall vacates a space for communication with nature. The structural design arrangement of this scenario is the center of the design. The void structure of lobby is nearly 20 meters long, and it is quite difficult to partition the entire lobby perfectly. In terms of structure, the designer boldly carried out the structural concept design of independent space and overhead floors. On the one hand, the sales objects were rationally and orderly diverted to ensure a certain degree of luxury. On the other hand, it guaranteed the order and privacy of sales. The design has vacated space from floor heights, which has allowed more room for functional division. The single display space like Party A's competitors is thus non-existed. A relatively smooth flow of movement between the office area and sample and sales room is thus created. It can also show the overall management capabilities of a large real estate company and will be a plus for later sales. The color tone is mainly gray, since after field investigation, it was found that the large-scale overall appearance of Party A's commercial podium is glass curtain wall and invisible LED screen. Taking account of the local location, lighting and temperature, the designer decided to equip the entire space with hard and cold feeling by using colors, which not only ensures the reduction of HVAC equipment investment, but also maximizes energy saving.

海棠餐厅

Begonia Canteen

虚实是中式设计重视的一个观念，典型的中式空间讲究"隔断"，这种隔断的目的并不在于要把空间切断，而是一个过渡、一种提醒、一种指示，常常"隔而不断"。碧纱橱、屏风、博古架、帷幕，不但"隔而不断"，还有很强的装饰性。譬如如今流行的造型墙、活动墙之类的元素，很久以前就被中国人运用自如了；又如墙面的留白、家具的虚功能和实体形的设计等，都是传统中式设计里的精髓。

Virtual-reality is a concept that Chinese design attaches great importance to. Typical Chinese design emphasizes "partition". The purpose of this partition is not to cut off the space, but a transition, a reminder, and an instruction, thus "isolated but continuous". The porch, screens, antique shelves, and curtains are not only "continuously separated", but also very decorative. For example, elements such as the popular modeling walls and movable walls have been freely used by the Chinese long ago. The white space of the wall, the design of the virtual function and the physical form of the furniture, etc., are all the essence in the traditional Chinese design.

项目名称：海棠餐厅
项目设计：广东星艺装饰集团
项目地址：山东泰安
设计师：洪泉

ONE·一个

ONE-Only

人一定要爱着什么，恰似草木对光阴的钟情。生活本就是一餐一饭，一生专心做好一件事。川菜就是此店创始人毕生所追求和钟情的。历经了20年的打拼，此店已成为当地知名的小吃店。人总会变老，生活也会逼着我们去成为大人，店家希望自己初出茅庐的儿子能抛弃稚嫩，变得稳重。伴随着脱离传统、为青年人打造合适的就餐环境的初衷，店家开始了品牌青年店的筹备。其筹备理念是：愿所有的年轻人都有充分的忍耐去担当，有充分单纯的心去信仰。

People must love something, just like plants' preference for time. Life is about a meal and something worthy of patience. Sichuan cuisine is what the founder of this restaurant has pursued and loved all his life. After 20 years of hard work, this restaurant has become a well-known local snack bar. People grow old, and life will force the next generation to become adults. The owner hopes that his fledgling son can abandon his immaturity and become prudent. With the original intention of breaking away from tradition and creating a suitable dining environment for young people, the restaurant has made full preparation. The preparation philosophy is: may all young people have enough patience to undertake responsibility and have a pure heart to believe.

项目名称：ONE·一个
项目设计：广东星艺装饰集团
项目地址：重庆万州
设计师：贺雷

在水一方

By the Water

前望遇龙河，后见月亮山，绝佳的位置造就一方美。酒店处在遇龙河景区至美地段，被月亮山环抱着，造就了"结庐在人境，而无车马喧"的怡人风情，少长咸集，群贤毕至，和谐自然的桌椅赋予其更多的东方文化韵味。白日里，光线穿过口字形门窗，卧室视野通透，窗外风景一览无遗。夜幕下，月光洒满象牙白的屋顶，静穆而又美好。

Overlooking the Yulong River, backed by Moon Mountain, the perfect location creates a beautiful place. The hotel occupies the most beautiful area of Yulong River scenic spot, surrounded by Moon Mountain, creating a pleasant atmosphere of " built my lodge amongst the throng of men without noise of horse and carriage". The old and young gathered and all the persons of virtue arrived, sitting by the harmonious and natural chairs and tables, immersed in an oriental cultural charm. In the daytime, the light passes through the square doors and windows. The bedroom has a transparent view and the scenery outside the window is unobstructed. Under the cover of darkness, the moonlight sprinkles over the ivory roof, peaceful and beautiful.

项目名称：在水一方
项目设计：广东星艺装饰集团
项目地址：广西桂林
设计师：陆勇

都市森林·ROOM 26 Whisky Bar

Urban Forest-Room 26 Whisky Bar

酒吧的名字表达了两位年轻海归女老板开酒吧的初心，26 是她们共同的幸运数字，这里是她们下班后属于自己的朋友圈……

设计师把酒吧打造为华丽且神秘的场所，让客人仿佛置身于电影的幻想世界之中。酒吧的创意灵感源自都市森林，那种充满神秘感的活力激情在隐隐释放，让人仿佛于光影交错的热带丛林中开启了一段奇妙的冒险之旅，惊喜连连。

The name of the bar expresses the original intention of the two young female overseas returnees. 26 is their lucky number. The bar is also a place to meet new friends after work...

The designer makes the bar a gorgeous and mysterious place, making guests feel as if they are in the fantasy world of movies. The creative inspiration of the bar comes from the urban forest. The energetic passion full of mystery is faintly released, bringing people into a wonderful adventure in the tropical jungle of light and shadow.

项目名称：都市森林·ROOM 26 Whisky Bar
项目设计：广东星艺装饰集团
项目地址：天津
设计师：马呈龙

极具私密性的入口扰乱着人的思绪，是男装定制剪裁店，还是凹凸有致的巨型金色巧克力墙？进入后却给客人带来意想不到的体验。酒吧深灰色的天花板采用了极简的裸露式设计，与丝绒材质的座椅以及灰棕鱼骨拼成的木地面形成鲜明对比。作为核心功能区的吧台上还有一处颇有艺术性的设计——六组水滴吊灯，暗示鸡尾酒的调制过程，生动精妙。

The very private entrance disturbs people's thoughts. Is it a custom tailoring shop for men's clothing, or a giant golden chocolate wall with bumps? After entering, it brings guests an unexpected experience. The dark gray ceiling of the bar adopts a minimal exposed design, in sharp contrast with the velvet seats and the wooden floor made of gray-brown fish bones. As the core functional area, there is also a very artistic design on the bar counter—six groups of drop-like lights, suggesting the cocktail preparation process, vivid and delicate.

在灯具选择上，设计师打破以往餐厅的灯光陈设，全部采用暖光玻璃、金属元素吊灯及工业风的轨道灯，让每个空间都能呈现出统一且错落的光感，为客人打造最舒适的品酒环境。隔层部分，占有绝对空间优势的吧台以流线型的古木纹石材台面及金色拉丝不锈钢为主要元素，搭配绿色植物壁纸，将大厅和玄关自然地连接起来。

In the selection of lamps, the designer breaks through the previous lighting furnishings of the restaurant. Warm glass, metal element chandeliers and industrial style track lights are adopted, so that each space can present a unified and scattered light, creating the most comfortable environment for guests to taste wine. In the compartment part, the bar counter with absolute space advantage takes streamlined ancient wood grain stone counter top and golden brushed stainless steel as the main elements, matched with green plant wallpaper, naturally connecting the hall and the entrance.

碧桂园四月茶生活美学馆

April Tea Living Art Studio in Country Garden

佫大的茶室只有一处茶席、一套茶具、一束插花、几张木椅，简约到了极致，但绝不简单。禅茶一味的茶室气息吸引的不单单是客人迷茫不定的眼睛，还有那份浮躁难安的心绪。觅城市一隅，享片刻宁静，沐心入境。

项目名称：碧桂园四月茶生活美学馆
项目设计：广东星艺装饰集团
项目地址：贵州六盘水
设计师：柯冬冬

The huge tea studio has only one tea table, one set of tea sets, a bunch of flower and a few wooden chairs revealing extreme simplicity, but fairly complicated. The scent of Zen tea attracts not only the uncertain eyes of the guests, but also the impetuous mood. Seek a corner in the city, enjoy a moment of tranquility, and entertain your heart.

自在天地

Free World

推门而入，一声热情的问候，一杯暖心的茶水，是否让等候的你心怀期待？很多时候我们护理的不仅仅是一张脸，而是一种健康、优雅、安全、积极向上的生活方式。现代女性坚持不懈地走在美丽、家庭和事业的道路上，越来越自信。她们能在职场披荆斩棘无所畏惧，也能系上围裙精心准备一桌羹汤。

Are you looking forward to a warm greeting, a cup of heart-warming tea behind the door? For most of the time, what we care about is not just a face, but a healthy, elegant, safe and positive lifestyle. Modern women are unremittingly walking on the path of pursuing beauty, family and career, becoming more and more confident. They can be fearless in the workplace as well as carefully in the kitchen, wearing an apron to prepare a table of dishes elaborately.

项目名称：自在天地
项目设计：广东星艺装饰集团
项目地址：广东广州
设计师：林婷婷

休闲区演绎"烟火璀璨"的装饰艺术，手边是托底小圆桌，花艺摆件和台灯的陈设变化充盈着书香雅意。除了传统的美容室，还增添了独立的 SPA 区。护理皮肤后，再做个美美的 SPA 放松自己，可谓一场完美的精神享受。秉持对"美"与"品质"一贯的高要求，连接功能区域之间的过道也讲究质感，巨大的梳妆镜融入整个空间，微妙的变幻在光与影的世界里回旋，单是过道就已经很美，心情瞬间变好。

The leisure area interprets the decorative art of "dazzling fireworks ". There is a small round table with a supporting base on hand, and the changes in floral decorations and table lamps are full of scholarly elegance. In addition to the traditional beauty room, a separate spa area has been added. After the skin care, try to relax yourself with a SPA, which is a perfect spiritual enjoyment. Adhering to the consistent high pursuit for "beauty" and "quality", the aisle connecting the functional areas also concerns the texture. The huge vanity mirror is integrated into the entire space, and the subtle changes revolve in the light and shadow. The aisle alone is beautiful enough, which lightens the mood instantly.

永乐路国贸 001-101 威克斯酒吧

VICS Bar in International Trade Center on Yongle Road 001-101

本案位于凯里市国贸商业中心一层，原建筑形态为一面落地玻璃及三面实体墙，由于紧邻户外马路，阳光及景物毫无遮挡。但对于一个面积只有 160 平方米的酒吧而言，一览无余的空间场景无疑是无趣且乏味的。如何在保证使用功能明晰的前提下，让空间各界面有丰富的视觉观感和动静相宜的区域设置，成为本案的设计重点。

This scenario is located on the first floor of the International Trade and Commercial Center in Kaili City. The original building is dominated by a floor-to-ceiling glass and three solid walls. Since it is close to the outdoor road, the sun and scenery are not blocked. But for a bar with an area of only 160 square meters, a unobstructed space is undoubtedly boring and tedious. How to make each interface of the space have rich visual perception as well as dynamic and static regional settings under the premise of ensuring the clear function for use has become the design focus of this project.

项目名称：永乐路国贸 001-101 威克斯酒吧
项目设计：广东星艺装饰集团
项目地址：贵州凯里
设计师：李少林

设计师将空间分割成室外和室内两个大的功能区域，同时利用层高的优势设立私密夹层，在上面可以看到吧台前方的通高酒架，它界定出酒吧的前后场域。而它和吧台之间是客人和调酒师互动的场所，刻意抬高组合过的亨利·卢梭的绘画作品，在黑白的丛林中，威克斯酒吧色彩变幻的 logo 就在不经意间突现。

在这里，设计师希望每位客人都能找到适合自己的那个角落，望出去或看进来都能欣赏到不同景色，"移步换景"也就有了现代意义的表达。又或者像店名 VICS 这个英文单词释义一样：树篱，是围合的，也是束缚的；是对冲的，也是两面的。酒吧作为现代人释放、放松自我的场所，偶尔打开内心，兴许你能碰到另一个久违的自己。

The designer divides the space into two large functional areas, outdoor and indoor. At the same time, the advantage of the height of the floor has been used to set up a private mezzanine. On the top, the high wine rack in front of the bar is clearly viewed, which defines the front and rear areas of the bar. Between it and the bar counter is a place for interaction between guests and bartenders. The combined paintings of Henri Rousseau are deliberately elevated. In the black-and-white jungle, the colorful logo of VICS Bar emerges inadvertently.

Here, the designer hopes that every guest can find a corner that suits them, and they can enjoy different scenery when looking out or looking in, and thus the traditional technique of "shifting the scenery in steps" bears a modern expression. Or like the definition of the English word VICS: hedges are enclosed and bound, opposite but connected. As a place for modern people to release and relax themselves, the bar occasionally helps to open the heart to meet another long-lost self.

售楼处

Sales Center

本案不仅从空间进深的角度去打动消费者，而且从不同的角度去思考如何营造一个多维度、立体的空间。设计师通过对视觉、触觉、听觉以及嗅觉的分析，全方位地打造售楼处的每一个细节。空间以白色为主，局部点缀温暖的木色和绿色。色彩的和谐搭配使空间平添了一份干净利索和温暖。

This scenario not only impressed consumers from the perspective of space depth, but also considered how to create a multi-dimensional and three-dimensional space from different aspects. Through the analysis of sense of vision, touch, hearing and smell, the designer concerns every detail of the sales office in an all-round way. The space is dominated by white, with warm wood and green embellishment. The harmonious collocation of colors adds neatness and warmth to the space.

项目名称：售楼处
项目设计：广东星艺装饰集团
项目地址：辽宁葫芦岛
设计师：冷娱良

初探

Primary Exploration

本案命名为"初探"，因为业主与设计师第一次探索餐饮这个行业。业主长期在以色列活动，想要在自己的咖啡厅里面融入一些以色列文化风情。经过了解和深思，设计师把以色列特色遗迹——"哭墙"融入休闲区。

This scenario was named "Primary Exploration" since both the owner and the designer explored the catering industry for the first time. The owner has been active in Israel for a long time and wants to incorporate some Israeli culture in his coffee shop. After mutual understanding and contemplation, the designer integrated the Israeli characteristic relic—the "Wailing Wall" into the leisure area.

项目名称：初探
项目设计：广东星艺装饰集团
项目地址：广东广州
设计师：陈智超

本案整体空间是从小到大，为了最大限度地利用空间，设计师把使用量最大的吧台设计成了异形吧台，增强了空间的流线感，也使这个空间的使用面积最大化。而整体空间采用暖色调来营造温暖、舒适、放松的感觉。

本案定位舒适、温馨，空间地面采用两种材质——休闲区的木纹砖、散座区的地平漆。休闲区采用文化石来模拟以色列"哭墙"，以色列友人来此后评价很亲切，整体空间墙面采用清水漆模拟原始水泥。

The overall space of this scenario is from small to large. In order to maximize space, the most used bar counter is designed with a unique shape, which enhances the flow of space and maximizes the space. The overall space uses warm colors to create a feeling of warmth, comfort and relaxation.

The orientation of this scenario is comfortable and homey. The space floor utilizes two materials—wood blocks in the leisure area and floor paint in the scattered seating area. The leisure area uses cultural stones to simulate the "Wailing Wall" in Israel. Israeli friends have given very cordial comments to the design. The walls of the entire space are made of clear water paint to simulate the original cement.

轻·玥

Gentle-Pearl

本案的设计注重给大众营造一个舒适、放松的氛围，设计风格是当下流行的轻奢风，整个空间使用了一些金属材质，包括内嵌灯管。色彩上以爱马仕橙、Tiffany 蓝为主色调，贯穿整个空间。利用原建筑四米多的层高，做了一些金属线灯，以及改良设计版的水泥树。从东门通过小的门厅进入，正厅是一处宽敞的前台。前台的柜面是融入的生态鱼缸设计，让客人在结账的同时还能欣赏到鲜活游动的鱼。一楼的格局是以厨房及前台为中心的可环绕式设计，可往左走，亦可往右走，左右各有通往二楼的楼梯，还设计了一个对外的小奶茶店，因为地处区域商业街，里面的包厢考虑设计为大包厢。

The design of this scenario focuses on creating a comfortable and relaxing atmosphere for the public. The design style is the popular mild luxury style. The entire space uses some metal materials, including embedded lamps. The color is mainly Hermès orange and Tiffany blue, which runs through the entire space as theme color. Some metal wire lights and the renovated design of the cement tree are installed by making full use of the 4-meter floor height of the original building. Through a small foyer from the east gate, the main hall is encountered with a large front desk. The counter at the front desk is an integrated ecological fish tank, allowing guests to appreciate the swimming fish while checking out. The entire layout of the first floor is a wrap-around design centered on the kitchen and the front desk. Either way, to left or to right is possible. There are two staircases leading to the second floor on both the left and right side. A small milk tea shop is also designed for the customers. Because it is located on a regional commercial street, the loge is considered to be designed as a large one.

项目名称：轻·玥
项目设计：广东星艺装饰集团
项目地址：江苏常州
设计师：方芳

　　本案周边有一些写字楼，如若有小团队可以在这里开会、聚餐。半开敞的隔断全部采用的是铁艺结合木地板的形式。楼梯处配置了一个网红翅膀，满足大小朋友的天使心。南楼梯往上是一处休闲区，顶上是仿佛满天星的灯。二楼以棋牌室为主，客人可以在这里接电话或者抽支烟，抑或随手拿起本感兴趣的书，安静地休息一会儿。每个包厢都有不同的颜色和风格。

There are some office buildings around. Meetings and dinners here are suitable for a small team. The semi-open partitions all adopt the form of wrought iron combined with wooden floor. There is a wing decoration, which is popular online, on the stairs to satisfy the angel hearts of both the old and young. Up the south staircase is a recreation area, with a star-filled lamp on top. The second floor is dominated by the chess and card room, where you can answer the phone or smoke a cigarette as well as pick up a book of interest and take a quiet rest. Each loge is equipped with different colors and styles.

茶生活

Tea and Life

Tea and Life expresses the proprietor's yearning for a paradise-style life, that people taste tea and wine, and spend a moment of leisure. In the hundreds of thousands of times in life, the ups and downs are just like tea leaves in a tea pot. A cup of tea encompasses every taste of life. Either clear and sweet or bitter and throat-cutting... Heating cups, setting, brewing, pouring, serving and tasting tea, pretense and impetuousness are thus removed step by step.

On walking into the tea house, you can slow down, calm down, and feel the poetry and beauty in life, where you can breathe freely under high pressure. Surrounded with the fragrance of tea, you can experience the charm of "slow" life and the joy of flow. The owner and designer combine the feelings of ancient literati life with the daily habits of modern life to create a new Chinese-style living aesthetic environment. The materials used in the space are wood, bamboo, rattan, brown, lacquer, and stone with soft and delicate colors, reflecting a simple beauty in the natural atmosphere and real texture. Strictly speaking, Qiwan Tea House is not a simple tea house, but a comprehensive life experience hall with new Chinese aesthetics. It integrates a humanistic tea house, private gourmet food, and a Chinese elegant collection. It contains a sense of quaint and seclusion. When I was here, an air of elegance shook my heart. Here you can wash your body and mind with peace of mind, remove the "burden" and relax yourself. Tasting tea, smelling fragrance, eating noodles, appreciating the taste, drinking wine... It's very happy to meet three or five friends and chat with you. The quiet, low-key and elegant place can let you put down your fatigue, feel the gentle time quietly, and reminisce the past prosperous years.

茶生活表达了业主对世外桃源式生活的向往，品茶、对酒，度一时清闲。人生百转千回间，浮沉翻滚犹如泡在茶杯中的茶叶。一杯茶，便囊括了百味人生。或清冽甘甜，或苦涩割喉……温具、置茶、冲泡、倒茶、奉茶、品茶，一步步去掉矫饰，赶走浮躁。

当你走进茶堂，便可以让自己慢下来、静下来，感受生活中的诗意与美好。在高压下找到一个可以自由呼吸的空间，在茶香中，体会"慢"生活的魅力，静品流动的喜悦。业主与设计师结合古时文人生活的情怀与现代生活的起居习惯，创造出新中式生活美学环境。空间的选材都是色泽柔和细腻的木、竹、藤、棕、漆、石，在自然的气息与真实的质感中体现一种朴素之美。七碗茶堂，从严格意义上说并不是简单的茶室，而是一家新中式美学的综合生活体验馆，集人文茶室、私品美食、中式雅集为一体，蕴含古雅之意与隐逸之风。置身于此，一股雅气荡然心中。在这里你能安心洗涤身心，卸下"包袱"，放松自我。品茶、闻香、吃面、赏味、对酒……约三五好友，寄情闲谈，甚是快哉！这里闹中取静、低调优雅，可以让你放下一身的疲惫，静静感受温柔时光，细细回味峥嵘岁月。

项目名称：茶生活
项目设计：广东星艺装饰集团
项目地址：广西贵港
设计师：周钦、孙章婷

时·光

Time-Tunnel

本案位于某商业城，整个项目设计延续原有办公室风格，并结合部门的使用需求进行整体空间划分。为更好地连通上下层区域，前台空间设计是本案的设计重点，设计师巧妙地将大理石与钢板进行结合，搭建一条通往二楼的时空隧道，行走在隧道中，每个转角都有不同的视觉体验。

This scenario is located in a commercial hall. The design of the entire project continues the original office style, and the overall space is divided in accordance with the needs of the department. In order to better connect the upper and lower floors, the design of the front desk is the focus of this scenario. The designer cleverly combined marble and steel plate to build a space-time tunnel leading to the second floor. Walking in the tunnel, you can encounter different visual experience at every corner.

项目名称：时·光
项目设计：广东星艺装饰集团
项目地址：广东广州
设计师：郑杰勇

Health Care International

Health Care International

本案业主是一位年过半百的成功人士，经营养生馆已十载有余，对于弧形的物体情有独钟。本案主题定位为"品味奢华"，致力于打造与众不同、独特新颖的空间。

本案在空间处理上运用了"曲线"的设计手法，让大厅不单显大气，更让人产生视觉上的震撼；主体风格以典雅欧式为主，也结合了一些奢华的手法，给人一种高贵而不失风雅的感觉。格局的改动更加符合人性设计，安排合理，无论是视觉或心理感受都是协调的。

The owner of this scenario is a successful person about fifty years old. He has running the Health Center for more than ten years and is fond of curved objects. The theme of this scenario is "tasting luxury" and is committed to creating a distinctive, unique and novel space.

The "curve" design technique is used in the space, which makes the hall not only full of magnificent atmosphere, but also make people feel visually shocked. The style is mainly elegant European style, but also combines some luxurious techniques to give people a noble feeling. The change of the layout is more in line with the humanity. With reasonable arrangement, both the visual and psychological feelings are harmonious.

项目名称：Health Care International
项目设计：广东星艺装饰集团
项目地址：四川宜宾
设计师：邓宏明

4 公共·方案设计作品

Public-Scenario Design Works

匀质空间

Even Space

　　建筑唯一正确的目标是依照自然的法则而建造。除非有正当理由，否则不要做画蛇添足的事情，多余的部分会随着时间的推移变得越发丑陋。

　　自然，和谐，合理，平衡。

　　空间中最重要的地方，往往是最不重要的。

　　在重要的地方花了很多力气，反而将这个地方的审美做得重复和过重。其实只要把次要的地方做好，重要的地方就凸显出来了，这就是空间的平衡感。

The only correct goal of architecture is to build in accordance with the laws of nature. Unless there is a valid reason, there is no need for superfluous decorations, or they will become more and more ugly over time.

It is about nature, harmony, fair and balance.

The most important place in space is usually the least important.

If lots of effort were spent in important places, the aesthetics of this place would be repeated and excessive. In fact, as long as the secondary places are done well, the important places will be highlighted. This is the sense of space balance.

项目名称：匀质空间

项目设计：广东星艺装饰集团

项目地址：广西百色

设计师：李飞

空间的美感可以通过块面、体量感、穿插来表达，哪怕是单一材质。

作为设计师，在进行办公空间设计时，一开始就需要确定一个方向。在本案设计中，设计师致力于让工作者感受到美的存在。

The beauty of space can be expressed by block surface, perception, and alternation, even with a single material. As a designer, when designing an office space, a direction from the beginning is a must. In the design of this scenario, the designer dedicates to make the workers feel the existence of beauty.

千合

Qian He

本案的甲方为建筑设计公司，对办公环境的设计要求较高，同时对空间也有着自己的理解，希望工作空间是纯粹的、有设计感的。本案采用建筑设计语言，把盒子、块面作为设计的一种手法，使空间更富有设计感，运用朴素的材料，如乳胶漆、木饰面、玻璃等，使空间更为纯粹。

电梯厅、会议室的设计也以块面为主，以灰白亮色为主色调，力求营造令人轻松愉悦的办公环境。总经理办公室以满足功能性为主。

办公区以原顶为主，尽量保证建筑层高。深色顶面与白色吊顶形成强烈反差，让空间更富有层次感。餐厅包房在设计手法上与办公区域有所不同，采用现代轻奢风格，给宾客以尊贵感。

The Party A in this scenario is an architectural design company. They have high requirements for the office environment. At the same time, they have their own understanding of the space. They hope that the work space is pure and has a sense of design. This scenario adopts the architectural design language, using boxes and block surfaces as a design technique to make the space appear more design-oriented, and using simple materials such as latex paint, wood veneers, and glass to make the space more pure.
The design of the elevator hall and meeting room is also based on block surfaces, with gray and white bright colors as the main color tone, striving to create a relaxing and pleasant office environment. The general manager's office is mainly designed to meet the functional requirements.
The office area is dominated by the original roof to ensure the height of floor. The dark top surface and the white suspended ceiling form a strong contrast, making the space more layered. The dining room is different from the office area in design techniques, adopting modern mild luxury style, which gives guests a sense of dignity.

项目名称：千合
项目设计：广东星艺装饰集团
项目地址：江苏南京
设计师：刘家德

滇池印象

Impression of Dian Lake

本案从东方美学中汲取精华，以现代手法来演绎中式设计。设计师重点考量空间的实用性与色调延展性，把人与空间联结起来，以人为本，牵引着空间规划的动线，把生活美学融入空间设计。

This scenario draws the essence from Oriental aesthetics and interprets Chinese design with modern techniques. The designer focuses on the practicality of the space and the extensibility of the color tone, which could connect people and space, with people-oriented attitude, leading the movement of space planning, and integrating life aesthetics into space design.

项目名称：滇池印象
项目设计：广东星艺装饰集团
项目地址：云南昆明
设计师：熊卫星

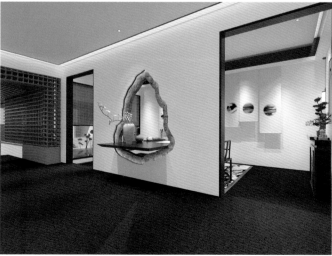

湘企城堡娱乐会所

Xiangqi Castle Club

本案为集大型空中舞台、空中特技表演场地、KTV 舞台情境体验为一体的娱乐会所，KTV 共设 39 间包房，有独立电梯直达每间包房，包房面积可达 200 平方米。每间包房的设计风格以时间为线索，以中国远古时期到近代的传奇人物为背景，充分结合人文性与娱乐性。

This scenario is an entertainment club that contains a large-scale aerial stage, performance venue of aerial acrobatics, and KTV stage. KTV has 39 private rooms, with independent elevators directly to each private room, and the private room area can reach 200 square meters. The design style of each private room takes time as a clue and is based on legends from ancient China to modern times, fully combining humanity and entertainment.

项目名称：湘企城堡娱乐会所
项目设计：广东星艺装饰集团
项目地址：贵州贵阳
设计师：李玉明

名宿酒店

Mingsu Hotel

名宿酒店与大海的直线距离为27米，有着超乎寻常的观海视觉效果。房间受原有建筑结构的局限，面积很小，设计师把精力放在了观景阳台的设计上，几何结构的阳台设计，让 20 世纪 70 年代的建筑焕发了新的生命力。

The linear distance between Mingsu Hotel and the sea is 27 meters. The hotel has an extraordinary visual effect. The room is limited by the original building structure. The designer thus puts his energy on the design of the viewing balcony. The geometric structure design irradiates the architecture of the 1970s with a lot of vitality.

项目名称：名宿酒店
项目设计：广东星艺装饰集团
项目地址：河北秦皇岛
设计师：刘天尧

渔村

Fishing Village

设计一家酒店，不是改变大自然，而是适应大自然。拥抱大自然，融入其中，人们能得到幸福和关怀。设计师通过手工竹丝灯、竹席、芦苇对空间进行了诠释，让人们在回味 20 世纪 80 年代的渔家往事中感受纯真年代的情怀故事。

To design a hotel is not to change nature, but to adapt to nature. Embracing nature and engaging with it, people can get happiness and care. The designer interprets the space through handmade bamboo silk lamps, bamboo mats, and reeds, allowing people to experience the feelings of the innocent age in the past of the fisherman in the 1980s.

项目名称：渔村
项目设计：广东星艺装饰集团
项目地址：广西桂林
设计师：陆勇

东方魅力

Oriental Charm

到处皆诗境，随时有物华，万物的起源皆来源于自然，建筑亦是其子。售楼处大堂中黑灰相间的大理石地板拟象，如湍流般将访客卷入山涧水喧；四周环形山水画取形，映出江南古韵，一呼一应之间塑造出现代与自然的离舍。洽谈区运用镜中世界的创意与借景手法呈现出多维度的广袤空间，视野开阔。褐色隐喻自然，与极具现代感的金属灯具搭配，凸显自由呼吸之感。

Poetic atmosphere and natural scenery are found everywhere. The origin of everything comes from nature, and so does the architecture. The simulacrum of the black and gray marble floor in the lobby of the sales office draws visitors into the mountains and rivers like a turbulent flow. The surrounding landscape paintings are shaped to reflect the ancient charm of the south of the Yangtze River, creating a modern and natural separation. The negotiation area uses the creativity of the world in the mirror and borrowing scenes to present a multi-dimensional vast space with a broad vision. Brown is a metaphor for nature, and it is matched with modern metal lamps to highlight the sense of free breathing.

项目名称：东方魅力
项目设计：广东星艺装饰集团
项目地址：广西钦州
设计师：陈振峰

奥特莱斯项目售楼处

Sales Office of Outlets Project

现代轻奢是近年比较流行的风格之一，低调奢华的设计让很多人为之着迷。现今社会，人们生活品质的提升，促使人们去追寻更高层次的享受——品位和高贵并存、舒适而优雅的生活，这就是轻奢风格的设计理念。在设计中，最常见的是金属、大理石、皮革制品，这种质感十足的选材能完美地凸显低调奢华，在舒适度与观赏度上，又能完全达到现代奢华峰点的要求。

Modern mild luxury is one of the most popular styles in recent years, and the low-key luxury design fascinates many people. In today's society, the improvement of people's quality of life prompts people to pursue a higher level of enjoyment—taste and nobility as well as comfortable and elegant life, which is the design concept of mild luxury style. In the design, metal, marble, and leather products are commonly used. This kind of texture selection can perfectly highlight the low-key luxury. In terms of comfort and view, it can fully meet the requirements of modern luxury.

项目名称：奥特莱斯项目售楼处
项目设计：广东星艺装饰集团
项目地址：山东济南
设计师：冷静

烟袅

Spiraling Smoke

轻烟徐徐回旋上升，若即若离，随风而逝。无论是何物，都如炊烟一样静静地来，静静地逝去，循环不已。夜色朦胧的村庄，在家家户户袅袅炊烟的映照下，更显一派人间自然景象。炊烟袅袅，意味着安静、和谐、温柔，表达了一种对自然的崇敬，以及把世间万物都看作一种自然常态的道家思想。

The light smoke slowly spirals and disappears with the wind, which seems so close but also so far. No matter what it is, it comes quietly like cooking smoke, passing away quietly, and circulating endlessly. In the dimly village, under the smoke from the households, the natural scene is even more harmonious. Spiraling smoke brings about a sense of peaceful, harmonious, and gentle. It expresses a kind of reverence for nature and regards everything in the world as a natural phenomenon like Taoism.

项目名称：烟袋
项目设计：广东星艺装饰集团
项目地址：广东广州
设计师：刘诚

一合料理

Yihe Cuisine

本案位于小区的商业街，定位是带有居酒屋功能的日料店。黑白灰加原木是整个店面的定调。为了不让顾客产生高冷的距离感，整体的灯光设计选用了温暖的暖光源，并针对居酒屋和料理店的切换设计了两套灯光照明系统。

本案主要是针对重点客户群体，研究他们的审美需求，最后呈现的设计方案也是以简约日式为主，利用日式设计的特点，用简约的手法结合结构设计表现出来。

This scenario is located on the commercial street of the community, and it is positioned as a Japanese food store with izakaya function. Black, white, gray and original wood are the tone of the entire store. In order not to keep customers from a high-cooling sense of distance, the overall lighting design uses a warm light source, and two sets of lighting systems are designed for the switch between the izakaya and the store. This scenario is mainly aimed at key customer groups, meeting their aesthetic needs. The final design plan is mainly simple Japanese style, using the characteristics of Japanese design and combining simple techniques with structural design.

项目名称：一合料理
项目设计：广东星艺装饰集团
项目地址：广西南宁
设计师：郭剑阳

蓝胜木业综合办公楼

Comprehensive Office Building of Lansheng Wood Industry

用极具戏剧性的设计语言体现企业的文化，充满张力和视觉冲击力的木制天花贯穿整个一楼展厅和洽谈接待区域。用简洁环保的生态板材把企业的产业和设计巧妙地融合到一起，既突出设计的巧妙构思，又展现企业低碳、节能、环保的经营理念。

The company's culture is reflected in a dramatic design language. The wooden ceiling full of tension and visual impact occupies the exhibition hall and reception area on the first floor. The company's industry and design are cleverly integrated with simple and environmentally friendly materials, which not only highlights the ingenious design concept, but also shows the company's low-carbon, energy-saving and environmentally friendly business philosophy.

项目名称：蓝胜木业综合办公楼
项目设计：广东星艺装饰集团
项目地址：广西贵港
设计师：蒋毅

兰章生活馆

Lanzhang Living Museum

本案是连排别墅，是集一层商业、二层道场、三层工作室于一体的个人品牌生活馆。以宋朝美学为空间氛围基调，也以佛家禅语"眼、耳、喉、鼻、舌、身、意"贯穿全案。一层入世，二层出世。二层道场，用于禅修，也会定期开坛授课。在设计空间上将老旧的木头作为设计元素，使空间的视重完全暗下来，只留了一点聚光。人同自己、同空间的对话寂静无声。

This scenario is a townhouse, a personal branded living hall that integrates a commercial area on the first floor, a dojo on the second floor and a studio on the third floor. Song Dynasty aesthetics is the keynote of the space atmosphere, and the Buddhist Zen language "eyes, ears, throat, nose, tongue, body, and mind" runs through the whole space. The first floor connects the world and the second floor an isolated one. The dojo on the second floor is used for meditation and regular classes. The old wood is used as a design element, so that the view of the space is completely darkened, leaving only a little spotlight. Thus, the dialogue between people and themselves and the space is silent.

项目名称：兰章生活馆
项目设计：广东星艺装饰集团
项目地址：广东佛山
设计师：夏芳芳

楚雄环洲企业办公楼

Headquarters Office Building of Chuxiong Huanzhou

　　本案是一个高端智慧办公大楼，集办公、休闲、接待、用餐为一体，构建高端商务价值体系。设计师考虑到对空间功能的需求，空间规划上重点考虑洽谈区域与办公区域的结合，做到动静分区有序，相对独立但又便于交流。

This scenario is a high-end smart office building that integrates office, leisure, reception, and dining areas to build a high-end business value system. Concerning the needs of space functions, the designer focuses on the combination of the negotiation area and the office area to ensure the orderly dynamic and static partitions which are relatively independent but easy to communicate.

项目名称：楚雄环洲企业办公楼
项目设计：广东星艺装饰集团
项目地址：云南楚雄
设计师：陈华、廖杰

案·序

Case-Order

设计师对空间进行重组，重点强化空间整体的流动性，利用材质延伸空间的视觉效果，强化室内建筑的概念，利用高度的变化及区域的围塑感，在原有的室内空间上重组建筑墙体的元素，利用挑空的手法增强空间的律动感。在整体色调及光线的安排上，使用深色的木饰面与水磨石，呈现沉稳及内敛的视觉效果。

The designer has reorganized the space, focusing on strengthening the overall fluidity of the space, using materials to extend the visual effect of the space. The concept of interior architecture is strengthened, using height changes and regional surroundings to reorganize the building walls on the original elements of interior space. Meanwhile, the void structure is used to enhance the rhythm of the space. In the overall arrangement of tone and light, dark wood veneer and terrazzo are used to present a calm and restrained visual effect.

项目名称：案·序
项目设计：广东星艺装饰集团
项目地址：广东广州
设计师：胡斌

梵净山度假酒店

Fanjingshan Resort Hotel

本案色调最终确定为木色与黑白色，禅文化与自然相对应，回归本真、素简。

结构上采用大面积的落地窗，意将大自然景色与室内融为一体，以求保持原始的本真本色，这是禅宗简素精神的一种体现。

从室内空间关系的层面来看，黑白是一个基础，是规矩；竹木是变化身形、恣意流动或瞬尔驻足、不逾矩的姿态。两者所呈现的形式正是在有无之间。具有自然象征意义的竹木通过大面积玻璃与户外的环境达成衔接，视觉引导心境，在"虚实""有无"间轻盈地结合。

The color of this scenario is determined to be wood color and black and white. Zen culture corresponds to nature, returning to the origin and simplicity.

The structure adopts a large area of floor-to-ceiling windows, which is intended to integrate the natural scenery with the interior, in order to maintain the original authenticity, which is a manifestation of the plain spirit of Zen.

From the perspective of the connection of indoor space, black and white are the basis and rules. Bamboo and wood present a stance of changing shapes, free flow, instant halt and ritual behavior. The form presented by the two is just between existence and nonexistence. The natural symbolic bamboo and wood are connected with the outdoor environment through a large area of glass, which enables the view to guide the mood and swiftly combining the "virtual and real" and "existence and nonexistence".

项目名称：梵净山度假酒店
项目设计：广东星艺装饰集团
项目地址：贵州铜仁
设计师：王少真

广聚传媒

Guangju Media

　　本案办公室设计手法与理念在于简单和实用，强调的是舍弃没有必要的装饰，用简洁的造型、愉悦的色彩来营造简洁、美观的效果。这种简约的设计非常适合当下快节奏的社会需求，人们在办公室办公，最需要的就是专注之余能感受氛围的舒适与轻松，在这种环境下工作正好可以放松人们上班紧绷的神经。

The design techniques and concepts of the office in this scenario are simplicity and practicality. The emphasis is on discarding unnecessary decorations and using simple shapes and pleasant colors to create simple and beautiful effects. This simple design is very suitable for the current fast-paced social needs. What people need most in the office is to feel the comfort and relaxation of the atmosphere while working. Working in this environment can just relax people's nervous mood at work.

项目名称：广聚传媒
项目设计：广东星艺装饰集团
项目地址：广西南宁
设计师：黄林妮

耕艺种德

设计幸福